PENGUIN BOOKS

A MATCH TO THE HEART

Gretel Ehrlich is the author of *The Solace of Open Spaces*, the novel *Heart Mountain*, and a collection of essays, *Islands, the Universe, and Home*.

Praise for *A Match to the Heart*

"Harrowing, amusing, unforgettably vivid . . . Ehrlich is a compulsive connector of the physical to the mystical. . . . Her contemplation of storms—their anatomy, behavior and power—is gripping, as she ponders the scale of what befell her."
—*Chicago Tribune*

"*A Match to the Heart* is opulent stuff . . . wonderfully evocative. . . . [Ehrlich] writes movingly of . . . the world with the vision of a mystic and the ecological consciousness of a naturalist."
—*New York Newsday*

"*A Match to the Heart* is in many ways the ultimate study of humans and nature, when what's outside suddenly forces its way inside."
—*The Sunday Oregonian*

"Like the title suggests, [Ehrlich] holds up a light to illuminate her own condition. The Buddhist in her gives us a thoughtful meditation on the state of suspension between life and death. The reporter in her shines a light on the science, myth and medicine of lightning."
—*Rocky Mountain News*

a match to the heart

Gretel Ehrlich

PENGUIN BOOKS

PENGUIN BOOKS
Published by the Penguin Group
Penguin Books USA Inc., 375 Hudson Street, New York, New York 10014, U.S.A.
Penguin Books Ltd, 27 Wrights Lane, London W8 5TZ, England
Penguin Books Australia Ltd, Ringwood, Victoria, Australia
Penguin Books Canada Ltd, 10 Alcorn Avenue, Toronto, Ontario, Canada M4V 3B2
Penguin Books (N.Z.) Ltd, 182-190 Wairau Road, Auckland 10, New Zealand

Penguin Books Ltd, Registered Offices: Harmondsworth, Middlesex, England

First published in the United States of America by Pantheon Books,
a division of Random House, Inc., 1994
Published in Penguin Books 1995

10 9 8 7 6 5 4 3 2 1

Grateful acknowledgment is made to the following for permission
to reprint previously published material:

Farrar, Straus & Giroux, Inc. and Faber and Faber Limited: Excerpt from "The Birthplace" from *Station Island* by Seamus Heaney, copyright © 1985 by Seamus Heaney. Rights outside the U.S. administered by Faber and Faber Limited, London. Reprinted by permission of Farrar, Straus & Giroux,
Inc. and Faber and Faber Limited.

New Directions Publishing Corp.: Excerpt from "Letter of Testimony"
from *A Tree Within* by Octavio Paz, copyright © 1988 by Octavio Paz.
Reprinted by permission of New Directions Publishing Corp.

THE LIBRARY OF CONGRESS HAS CATALOGUED THE HARDCOVER AS FOLLOWS:
A match to the heart/Gretel Ehrlich.
p. cm.
ISBN 0-679-42550-0 (hc.)
ISBN 0 14 01.7937 2 (pbk.)
1. Ehrlich, Gretel—Health. 2. Electrical injuries—Patients—Biography.
3. Lightning—Health aspects—Case studies. I. Title.
RD96.5.E37 1994
617.1′22—dc20
[B]

Printed in the United States of America
Set in Transitional
Designed by Fearn Cutler

ACKNOWLEDGMENTS

I want to thank all those who gave help, care, and friendship in a time of need. Heartfelt thanks to my parents, Grant and Gretchen Ehrlich, who saved me, cared for me, and gave moral support and new friendship in our late years; to Dr. Blaine Braniff, healer, teacher, and companion extraordinaire; to Dick and Dorothy Roberts for making their plane available; to the doctors and nurses at Santa Barbara Cardiovascular Group; to the nurses in the Coronary Care Unit at Cottage Hospital; to Dr. Mary Cooper whose medical literature on lightning injury has helped us all; to Dr. Rick Westerman, thoracic surgeon, who let me watch; and to all those at the Lightning Strike and Electric Shock conference who told me their stories. Thanks to friends, old and new, whose cheer brightened long nights and days, especially Sue Davies and David Buckland; Patrick Markey, Bob Redford, Theresa Curtain, and Carol Fontana; Tamara Asseyev, Noel and Judy Young, Hillary Hauser and Jim Marshall, Kate and Clyde Packer, Jim Cresson, Marshall and Heidi Rose, Laurel Miller, Aaron Young, Pico Iyer, and Michael Ross; to Naomi Seeger for help with medical research; to my agent, Liz Darhansoff; to my various editors and publishers: Michael Jacobs, Al Silverman, Paul Slovak, and my editor, Dan Frank; and, last but not least, my canine friends and guides, Rusty, French Fry, Yaki, and, especially, Sam.

Everywhere being nowhere,
who can prove
one place more than another?
We come back emptied,
to nourish and resist
the words of coming to rest:
birthplace, roofbeam, whitewash,
flagstone, hearth
like unstacked iron weights
afloat among galaxies.

—Seamus Heaney

. . . a witch who escapes human
detection will nevertheless
eventually be struck down by
lightning.

—Clyde Kluckhohn, *Navaho Witchcraft* (1944)

chapter 1

Deep in an ocean. I am suspended motionless. The water is gray. That's all there is, and before that? My arms are held out straight, cruciate, my head and legs hang limp. Nothing moves. Brown kelp lies flat in mud and fish are buried in liquid clouds of dust. There are no shadows or sounds. Should there be? I don't know if I am alive, but if not, how do I know I am dead? My body is leaden, heavier than gravity. Gravity is done with me. No more sinking and rising or bobbing in currents. There is a terrible feeling of oppression with no oppressor. I try to lodge my mind against some boundary, some reference point, but the continent of the body dissolves . . .

A single heartbeat stirs gray water. Blue trickles in, just a tiny stream. Then a long silence.

Another heartbeat. This one is louder, as if amplified. Sound takes a shape: it is a snowplow moving grayness aside like a heavy snowdrift. I can't tell if I'm moving, but more blue water flows in. Seaweed begins to undulate, then a whole kelp forest rises from the ocean floor. A fish swims past and looks at me. Another heartbeat drives through dead water, and another, until I am surrounded by blue.

Sun shines above all this. There is no pattern to the way its glint comes free and falls in long knives of light. My two beloved dogs appear. They flank me like tiny rockets, their fur pressed against my ribs. A leather harness holds us all together. The dogs climb toward light, pulling me upward at a slant from the sea.

I have been struck by lightning and I am alive.

chapter 2

Before electricity carved its blue path toward me, before the negative charge shot down from cloud to ground, before "streamers" jumped the positive charge back up from ground to cloud, before air expanded and contracted producing loud pressure pulses I could not hear because I was already dead, I had been walking.

When I started out on foot that August afternoon, the thunderstorm was blowing in fast. On the face of the mountain, a mile ahead, hard westerly gusts and sudden updrafts collided, pulling black clouds apart. Yet the storm looked harmless. When a distant thunderclap scared the dogs, I called them to my side and rubbed their ears: "Don't worry, you're okay as long as you're with me."

I woke in a pool of blood, lying on my stomach some distance from where I should have been, flung at an odd angle to one side of the dirt path. The whole sky had grown dark. Was it evening, and if so, which one? How many minutes or hours had elapsed since I lost consciousness, and where were the dogs? I tried to call out to them but my voice didn't work. The muscles in my throat were paralyzed and I couldn't swallow. Were the dogs dead? Everything was terribly wrong: I had trouble seeing, talking, breathing, and I couldn't move my legs or right arm. Nothing remained in my memory—no sounds, flashes, smells, no warnings of any kind. Had I been shot in the back? Had I suffered a stroke or heart attack? These thoughts were dark pools in sand.

The sky was black. Was this a storm in the middle of the day or was it night with a storm traveling through? When thunder exploded over me, I knew I had been hit by lightning.

The pain in my chest intensified and every muscle in my body ached. I was quite sure I was dying. What was it one should do or think or know? I tried to recall the Buddhist instruction regarding dying—which position to lie in, which direction to face. Did the "Lion's position" taken by the Buddha mean lying on the left or the right? And which sutra to sing? Oh yes, the Heart Sutra . . . gaté, gaté, paragaté . . . form and formlessness. Paradox and cosmic jokes. Surviving after trying to die "properly" would be truly funny, but the chances of that seemed slim.

Other words drifted in: how the "gateless barrier" was the gate through which one passes to reach enlightenment. Yet if there was no gate, how did one pass through? Above me, high on the hill, was the gate on the ranch that lead nowhere, a gate I had mused about often. Now its presence made me smile. Even when I thought I had no aspirations for enlightenment, too much effort in that direction was being expended. How could I learn to slide, yet remain aware?

To be struck by lightning: what a way to get enlightened.

That would be the joke if I survived. It seemed important to remember jokes. My thinking did not seem connected to the inert body that was in such terrible pain. Sweep the mind of weeds, I kept telling myself—that's what years of Buddhist practice had taught me. . . . But where were the dogs, the two precious ones I had watched being born and had raised in such intimacy and trust? I wanted them with me. I wanted them to save me again.

It started to rain. Every time a drop hit bare skin there was an explosion of pain. Blood crusted my left eye. I touched my good hand to my heart, which was beating wildly, erratically. My chest was numb, as if it had been sprayed with novocaine. No feeling of peace filled me. Death was a bleakness, a grayness about which it was impossible to be curious or relieved. I loved those dogs and hoped they weren't badly hurt. If I didn't die soon, how many days would pass before we were found, and when would the scavengers come? The sky was dark, or was that the way life flew out of the body, in a long tube with no light at the end? I lay on the cold ground waiting. The mountain was purple, and sage stirred against my face. I knew I had to give up all this, then my own body and all my thinking. Once more I lifted my head to look for the dogs but, unable to see them, I twisted myself until I faced east and tried to let go of all desire.

When my eyes opened again I knew I wasn't dead. Images from World War II movies filled my head: of wounded soldiers dragging themselves across a field, and if I could have laughed—that is, made my face work into a smile and get sounds to discharge from my throat—I would have. God, it would have been good to laugh. Instead, I considered my options: either lie there and wait for someone to find me—how many days or weeks would that take?—or somehow get back to the house. I calmly assessed what might be wrong with me—stroke, cerebral hemorrhage, gunshot wound—but it was bigger than I could understand. The instinct to survive does not rise from particulars; a deep but general misery rollercoasted me into action. I tried to propel myself on my elbows but my right arm didn't work. The wind had swung around and was blowing in from the east. It was still a dry storm with only sputtering rain, but when I raised myself up, lightning fingered the entire sky.

It is not true that lightning never strikes the same place twice. I had entered a shower of sparks and furious brightness and, worried that I might be struck again, watched as lightning touched down all around me. Years before, in the high country, I'd been hit by lightning: an electrical charge had rolled down an open meadow during a fearsome thunderstorm, surged up the legs of my horse, coursed through me, and bounced a big spark off the top of my head. To be struck again—and this time it was a direct hit—what did it mean?

The feeling had begun to come back into my legs and after many awkward attempts, I stood. To walk meant lifting each leg up by the thigh, moving it forward with my hands, setting it down. The earth felt like a peach that had split open in the middle; one side moved up while the other side moved down and my legs were out of rhythm. The ground rolled the way it does during an earthquake and the sky was tattered book pages waving in different directions. Was the ground liquifying under me, or had the molecular composition of my body deliquesced? I struggled to piece together fragments. Then it occurred to me that my brain was torn and that's where the blood had come from.

I walked. Sometimes my limbs held me, sometimes they didn't. I don't know how many times I fell but it didn't matter because I was making slow progress toward home.

Home—the ranch house—was about a quarter of a mile away. I don't remember much about getting there. My concentration went into making my legs work. The storm was strong. All the way across the basin, lightning lifted parts of mountains and sky into yellow refulgence and dropped them again, only to lift others. The inside of my eyelids turned gold and I could see the dark outlines of things through them. At the bottom of the hill I opened the door to my pickup and blew the horn with the idea that someone might hear me. No one came. My head had swollen to an indelicate shape. I tried to swallow—I was so thirsty—but the muscles in my throat were still paralyzed and I wondered when I would no longer be able to breathe.

Inside the house, sounds began to come out of me. I was doing crazy things, ripping my hiking boots off because the bottoms of my feet were burning, picking up the phone when I was finally able to scream. One of those times, someone happened to be on the line. I was screaming incoherently for help. My last conscious act was to dial 911.

Dark again. Pressing against sore ribs, my dogs pulled me out of the abyss, pulled and pulled. I smelled straw. My face was on tatami. I opened my eyes, looked up, and saw neighbors. Had they come for my funeral? The phone rang and I heard someone give directions to the ambulance driver, who was lost. A "first responder," an EMT from town who has a reputation with the girls, leaned down and asked if he could "touch me" to see if there were any broken bones. What the hell, I thought. I was going to die anyway. Let him have his feel. But his touch was gentle and professional, and I was grateful.

I slipped back into unconsciousness and when I woke again two EMTs were listening to my heart. I asked them to look for my dogs but they wouldn't leave me. Someone else in the room went outside and found Sam and Yaki curled up on the porch, frightened but alive. Now I could rest. I felt the medics jabbing needles into the top of my hands, trying unsuccessfully to get IVs started, then strapping me onto a backboard and carrying me out the front door of the house, down steps, into lightning and rain, into what was now a full-blown storm.

The ambulance rocked and slid, slamming my bruised body against the metal rails of the gurney. Every muscle was in violent spasm and there was a place on my back near the heart that burned. I heard myself yell in pain. Finally the EMTs rolled up towels and blankets and wedged them against my arms, shoulders, hips, and knees so the jolting of the vehicle wouldn't dislodge me. The ambulance slid down into ditches, struggled out, bumped from one deep rut to another. I asked to be taken to the hospital in Cody, but they said they were afraid my heart might stop again. As it was, the local hospital was thirty-five miles away, ten of them dirt, and the trip took more than an hour.

Our arrival seemed a portent of disaster—and an occasion for comedy. I had been struck by lightning around five in the afternoon. It was now 9:00 P.M. Nothing at the hospital worked. Their one EKG machine was nonfunctional, and jokingly the nurses blamed it on me. "Honey, you've got too much electricity in your body," one of them told me. Needles were jammed into my hand— no one had gotten an IV going yet—and the doctor on call hadn't arrived, though half an hour had elapsed. The EMTs kept assuring me: "Don't worry, we won't leave you here." When another nurse, who was filling out an admission form, asked me how tall I was, I answered: "Too short to be struck by lightning."

"Electrical injury often results in ventricular fibrillation and injury to the medullary centers of the brain. Immediately after electric shock patients are usually comatose, apneic, and in circulatory collapse. . . ."

When the doctor on call—the only doctor in town, waddled into what they called the emergency room, my aura, he said, was yellow and gray—a soul in transition. I knew that he had gone to medical school but had never completed a residency and had been barred from ER or ICU work in the hospitals of Florida, where he had lived previously. Yet I was lucky. Florida has many lightning victims, and unlike the doctors I would see later, he at least recognized the symptoms of a lightning strike. The tally sheet read this way: I had suffered a hit by lightning which caused ventricular fibrillation—cardiac arrest—though luckily my heart started beating again. Violent contractions of muscles when one is hit often causes the body to fly through the air: I was flung far and hit hard on my left side, which may have caused my heart to start again, but along with that fortuitous side effect, I sustained a concussion, broken ribs, a possible broken jaw, and lacerations above the eye. The paralysis below my waist and up through the chest and throat—called kerauno-paralysis—is common in lightning strikes and almost always temporary, but my right arm continued to be almost useless. Fernlike burns—arborescent erythema—covered my entire body. These occur when the electrical charge follows tracings of moisture on the skin—rain or sweat—thus the spidery red lines.

"Rapid institution of fluid and electrolyte therapy is essential with guidelines being the patient's urine output, hematocrit, osmolality, central venous pressure, and arterial blood gases. . . ."

The nurses loaded me onto a gurney. As they wheeled me down the hall to my room, a front wheel fell off and I was slammed into the wall. Once I was in bed, the deep muscle aches continued, as did the chest pains. Later, friends came to visit. Neither doctor nor nurse had cleaned the cuts on my head, so Laura, who had herded sheep and cowboyed on all the ranches where I had lived and whose wounds I had cleaned when my saddle horse dragged her across a high mountain pasture, wiped blood and dirt from my face, arms, and hands with a cool towel and spooned yogurt into my mouth.

I was the only patient in the hospital. During the night,

sheet lightning inlaid the walls with cool gold. I felt like an ancient, mummified child who had been found on a rock ledge near our ranch: bound tightly, unable to move, my dead face tipped backwards toward the moon.

In the morning, my regular doctor, Ben, called from Massachusetts, where he was vacationing, with this advice: "Get yourself out of that hospital and go somewhere else, anywhere." I was too weak to sign myself out, but Julie, the young woman who had a summer job on our ranch, retrieved me in the afternoon. She helped me get dressed in the cutoffs and torn T-shirt I had been wearing, but there were no shoes, so, barefoot, I staggered into Ben's office, where a physician's assistant kindly cleansed the gashes in my head. Then I was taken home.

Another thunderstorm slammed against the mountains as I limped up the path to the house. Sam and Yaki took one look at me and ran. These dogs lived with me, slept with me, understood every word I said, and I was too sick to find them, console them—even if they would have let me.

The next day my husband, who had just come down from the mountains where he worked in the summer, took me to another hospital. I passed out in the admissions office, was loaded onto a gurney, and taken for a CAT scan. No one bothered to find out why I had lost consciousness. Later, in the emergency unit, the doctor argued that I might not have been struck by lightning at all, as if I had imagined the incident. "Maybe a meteor hit me," I said, a suggestion he pondered seriously. After a blood panel and a brief neurological exam, which I failed—I couldn't follow his finger with my eyes or walk a straight line—he promptly released me.

"Patients should be monitored electrocardiographically for at least 24 hours for significant arrhythmias which often have delayed onset. . . ."

It was difficult to know what was worse: being in a hospital where nothing worked and nobody cared, or being alone on an isolated ranch hundreds of miles from decent medical care.

In the morning I staggered into the kitchen. My husband, from whom I had been separated for three months, had left at 4:00 A.M. to buy cattle in another part of the state and would not be back for a month. Alone again, it was impossible to do much for myself. In the past I'd been bucked off, stiff and sore plenty of times but this felt different: I had no sense of equilibrium. My head hurt, every muscle in my body ached as if I had a triple dose of the flu, and my left eye was swollen shut and turning black and blue. Something moved in the middle of the kitchen floor. I was having difficulty seeing, but then I did see: a rattlesnake lay coiled in front of the stove. I reeled around and dove back into bed. Enough tests of character. I closed my eyes and half-slept. Later, when Julie came to the house, she found the snake and cut off its head with a shovel.

My only consolation was that the dogs came back. I had chest pains and all day Sam lay with his head against my heart. I cleaned a deep cut over Yaki's eye. It was half an inch deep but already healing. I couldn't tell if the dogs were sick or well, I was too miserable to know anything except that Death resided in the room: not as a human figure but as a dark fog rolling in, threatening to cover me; but the dogs stayed close and while my promise to keep them safe during a thunderstorm had proven fraudulent, their promise to keep me alive held good.

chapter 3

Days went by. When I took a bath the stench of burned hair and skin drove the dogs out of the house. The hot place on my back still felt as if an ember had been buried under the skin but no blister appeared. I lay on a narrow daybed in the library unable to climb the stairs to the bedroom. Friends who were shooting a film in Montana thought it unwise for me to be alone and sent rescuers: Theresa, Carol, and Marsha came to the ranch, packed my things, and drove me to Livingston, Montana, where I would be near a hospital and could be watched by the medic on location every day.

When we drove into the motel parking lot I experienced a blank: I could not figure out where I was or why. There was no memory of the transaction or logic that had brought me to that place and I wanted only to go to bed. From the car I rose stiffly, my ability to move only slightly better than my ability to comprehend. Inside my room there were flowers and a note of condolence and congratulations on being alive. I was reminded that I had been nearly dead.

The setting was idyllic. The best fishing guides in Montana were on hand to teach the actors the art and "religious act" of landing a dry fly on moving water. The director, sleek and handsome in his wetsuit, was directing from midstream; French chefs served up glamorous meals to the crew; but I was having chest pains and difficulty staying conscious. While talking to friends I dropped to the ground. When I opened my eyes Robert Redford was riding a handsome black horse toward me, waving, smiling, and asking if I wanted to go for a ride. *Is this what it's like being dead?*

In the emergency room at the hospital in Bozeman, nothing conclusive about my problems was found. After a few hours I was released. As I was walking out the door the ER nurse asked why I was shuffling and when I shrugged, she went back to reading her magazine.

During the week my headaches worsened but the neurologist in Billings wouldn't see me. Finally I insisted, but he declined to give me an EEG, saying they were "awfully expensive and I seemed fine." I met up with a rancher who had been hit by lightning years before. He said only, "You feel like hell for about three weeks, then you're okay again." So I carried on, expecting to feel better soon.

One afternoon a violent thunderstorm erupted over the film crew. We were standing by the Gallatin River under a tall stand of pines—the worst possible place to be because lightning was striking close. I still had gashes in my head, a black eye, cuts and bruises. There was no place to go to get out of danger. I tried to laugh at this sudden predicament but noticed no one was standing by me. They were all afraid that lightning would seek me out again. Finally Redford ran over, grabbed my hand, and pulled me up a steep hill into a stranger's house, where we were kindly given refuge.

I decided to return to the ranch. Once there I realized that my condition—whatever it was—had worsened. It was almost impossible for me to stay conscious. As I lay on my back with my feet up, the world grew black and a deadly lethargy filled me so that I could not move or talk. Chest pains, both sharp and piercing as well as deep and aching, with the classic heart attack symptoms—clamminess, shortness of breath, pains down the left arm—kept me awake all night.

In the morning I called my parents in California. A feeling had come over me that something was terribly wrong and I began to think I might not survive another day. The presence of death in the room was vivid. In mid-sentence I passed out and when I came to they were still on the phone. My father patiently asked if I thought I could get on a commercial flight and when I said no, I didn't think I could, he paused, then said, "I'll be there in four hours."

Lying on the floor of the ranch house, I felt the life in my body trying to ebb away. At times I couldn't even stand and, reaching up from the floor, pulled clothes off hangers and stuffed them into a duffel bag. At the time I had no idea I would never spend another night at the ranch.

A Beechcraft King Air dove down out of Wyoming's blue sky and landed at the Greybull airport. Somehow I had driven myself to the local landing strip. Nor was I so sick that I couldn't recognize a sexy plane. We boarded quickly, turned the plane around and headed for California. The pilot had brought sandwiches and sodas. With my feet propped up, I watched the Rocky Mountain cordillera give way to fins of red rock and the sacred mountains of the Navajo. High desert changed to Sonoran desert. The southern tip of the Sierras dwindled and melted into ground that stolen water had converted into cotton and alfalfa fields.

Half dead, I was being taken back to my childhood home. The desert behind us, I could now see the coastal range: ridges and ridges of blue, then brown grassland cut off abruptly by a green and turquoise sea. Floating above the edge of the continent, I felt I had been resurrected. Below, the Santa Ynez Mountains rose to Tranquillion Peak, then dwindled into a gently sloping hill that plunged into the blustery Pacific at Point Conception. This is the place the dead go, the local Chumash Indians say, as if the slope were a staircase of the dead. This is their gateway to the afterworld.

An automated lighthouse turned its head like an owl and winked its penetrating light into the plane. After death, according to Chumash legend, the soul first goes to Point Conception, which was a wild and stormy place called Humqaq. There, below a cliff that can only be reached by rope, is a pool of water into which water continually drips. In that stone basin can be seen the footprints of the dead. They bathe themselves there, then, seeing a light to the west, go toward it through the air, and reach the land of the dead, called Similaqsa.

I had risen from the dead, and to return home I had to pass over this desolate point, this charnel ground. The Chumash legend said that in the evening the people at a nearby village would see a soul passing on its way to Humqaq. They motioned with their hands at the soul and told it to return, and they clapped their hands. Sometimes the soul would respond and turn back, but other times it would simply swerve a little from its course and continue on to Similaqsa. Then it shone like a light and left a blue trail. Sometimes there was a fiery ball at its side, and as the soul passed by, there was a report like a distant cannon—or thunder—and the sound of a gate closing.

The plane turned south as the land here does, facing into the sun. Under us the jigsaw coast smoothed out and the sea churned, its white lights going on and off as if signaling us to land: HERE, HERE.

chapter 4

My mother had made an appointment with a prominent cardiologist kind enough to take me on an emergency basis. On the way to his office I began sinking into unconsciousness. By the time my mother pulled up in front of his building, I was gone.

Nurses and a doctor came out to the car. Semiconscious, I was wheeled into an examining room. But the darkness kept descending. Though unable to move, talk, or see, I could hear every utterance clearly. One of the doctors said, "I can't find a pulse."

Shirt off, EKG leads glued to my chest, I felt my mother stroke my hand. Then a doctor's rich-toned voice—that of Blaine Braniff, the cardiologist who would care for me—whispered, "I can't believe she's still alive."

Blood pressure and heart rate are supposed to compensate for each other: if the pressure goes down, the heart speeds up to pump more blood. But mine did the opposite: not only was a pulse difficult to find, but my heart was slowing down. I could hear the nurse reading the heart monitor: 50 . . . 40 . . . 30. . . . Is this how you die? I wondered. But every word being said rang in my ears with absolute clarity.

An IV with atropine was started, and very slowly, like a fly on a cold day, I began to come around. Once again, my two dogs—Sam and Yaki—seemed to be with me, pulling me out of that bleak state, sledding me up shafts of light to consciousness and safety. When I looked up, my doctor was standing over me. He had unarmored eyes I could look into, and before I was able to speak, he made me smile.

Long ago, when my friend David was dying of liver cancer, I dreamed he was trapped inside an opaque cocoon that kept rolling down a muddy hill. I could only see a faint outline of his emaciated body but could hear his keen voice: "I'm not dead yet," he said. Seventeen years later my own cocoon was black, a monk's cowl dropped over my brain, a diver's lead weight pulling me down into a gray sea, and no one could hear me.

Dark again, not black but gray, the vagueness I hated. But it felt good to lie still for as long as I wanted and rest. More dreams from David's death-days rang through me: David alive, but a skeleton, sitting in a wicker chair made of bones, beckoning me to come closer, laughing at the joke of existence which is the reality of death. Or lying on top of his coffin, one of many lined up around a circular driveway—not a skeleton this time but fully clothed, and, like the lighthouse at Point Conception, winking at me.

Dr. Braniff touched my forehead and hand. My father had arrived and was standing with my mother. I struggled to move and talk and sound normal, to reassure them—always the urge to appear better than one really feels. Is this training useful or not? I wondered. The nurses pulled the leads off my chest. "Okay, let's go," the doctor said, then drove me the single block to the hospital in his own car and, half-carrying me, passed through the emergency room straight up to the cardiac care unit.

I was hooked up and wired so that heart rate and blood pressure could be monitored. Blood-pressure cuffs were put on each arm, IV catheters inserted into my veins in case they were needed for atropine. In an acute care unit, male and female nurses work twelve-hour shifts and have only two or three patients apiece. They were attentive, kind, and curious, and the fishbowl effect of the rooms, with glass walls on three sides, increased the intimacy. Unlike the brutalizing effect of such close quarters in a jail, this elbow-rubbing was just what I needed. Death visited but was no longer ravenous, only a small figure stuck in the far corner of the room. I kept my eye on him. All around me were very sick people, some just back from open heart surgery, others with worn out hearts who were on ventilators and life-support machines. Yet no place could have seemed as welcoming to me. For the first time in three weeks I could relax.

The tests they gave me were the ones routinely given in the first hours after a lightning injury: another CAT scan to check for cerebral bleeding, X-rays for broken bones, blood panels to monitor kidney function and cardiac enzymes, EKGs for the late onset of arrhythmias and changes that might indicate tissue damage in the heart or lungs, and an overdue EEG to make sure my blackouts weren't epileptic seizures. Lying on a small bed with wires attached to my skull, I watched white printouts of brain waves stack up on a table beside me and wished the motions of mind, the hieroglyphics of imagination, were as accessible to me, that they were the pages of a novel coming into being.

While he read my books I studied the literature of lightning injury. My vision was blurred from the medication I was taking (Norpace), which made my heart beat regularly whether it wanted to or not. In a 1990 *Medical Journal of Australia* I read this: "Lightning as a cause of death and injury is a highly significant natural phenomenon with potentially devastating effects. Almost every organ system can be impaired when lightning, an electric current, passes through the human body taking the shortest path between the contact points. The overall mortality rate following lightning injury is 30% and in survivors the morbidity rate is 70%. In the United States there are several thousand lightning-related injuries reported every year with almost 600 deaths, a figure which makes lightning responsible for more deaths in the US than any other natural phenomenon." Wyoming, I learned, has the highest death rate per capita from lightning, but that's probably because any kind of medical care—good or bad—is hard to get to in such an underpopulated state.

During frequent visits to the cardiac care unit, Blaine demystified the medical jargon. Direct hits by lightning can cause unconsciousness and coma, cardiopulmonary arrest, or ventricular fibrillation, which is cardiac arrest, and autonomic nervous system damage. As millions of volts of electricity pass through the body, brain cells are burned, "insulted," or bruised, which can result in cerebral edema, hemorrhage, and epileptic seizures. Passing down through the body, electricity hits the soft tissue organs—heart, lungs, and kidneys—causing contusions, infarctions, coagulations, or cellular damage that can lead to death. Tympanic membranes in the ear sometimes burst from the explosion of thunder, and cataracts develop if the flash has been intensely bright. Cases of leukemia have been recorded, and when pregnant women are hit, either spontaneous abortion occurs, or else they carry the baby to full term but after delivery the infant dies. "It's no wonder you feel like hell for a while," I told him.

Later I read this: "Death from Lightning—the Possibility of Living Again" one case history announced. This is the seeming miracle for which lightning injury is famous. In 1961 a ten-year-old boy was struck while riding a bicycle. Slumped unconscious against a tree, the boy was given back-press, arm-lift artificial respiration, though on arrival at a hospital, twenty-two minutes later, he was apparently dead. For seven minutes no artificial respiration had been given. On admission his pupils were dilated, he was pulseless, and he was not breathing. Mouth-to-airway respiration was instituted immediately. When the chest was opened it did not bleed; the heart was motionless. Cardiac massage was given and epinephrine was injected into the left ventricle. Five minutes later heart action started and he was placed on a respirator. Twenty-nine days later he was discharged from the hospital, almost fully recovered. When he returned to school a month later, his IQ measured slightly higher than it did before the accident.

It is now thought that such miracles can happen because metabolic rates seem to come to a standstill after the body is hit by lightning—a hot-flash hibernation reaction, like the one induced by cold. Burns are the cause of another controversy. Why are some people burned and others not? When friends saw me after my accident they seemed disappointed. "I thought you'd have a black imprint of lightning across your face and down your shoulder and a streak of white hair, or else you'd be bald," one acquaintance told me.

Besides "Lichtenberg's flowers"—the transient, feathery pattern due to imprints from electron showers through the skin, not true burns at all—I had the one spot on my back that felt as if it were burning, though it left no mark. Whether someone is burned or not depends on the duration of the lightning stroke. "Cold lightning" is the first stroke, which carries much less heat and spends less time in the body than the "return stroke" from the ground back up to the cloud. Those hit by the return stroke are burned; those hit by the initial charge are not, and apparently I was one of those lucky ones.

A slow death, a stupid dwindling, or a fatal arrhythmia—that's what might have happened if I had stayed in Wyoming. My concussion and head wounds had healed, my balance was better, strength was returning to my limbs, my memory was good (though I didn't feel very fast or sharp), and the terrible aches and pains were subsiding. But what deeper injury was causing the problems that persisted, what kind of chemical and electrical chaos had been generated in my body when all those volts of electricity passed through?

The lower my blood pressure fell, the slower my heart

beat. Insufficient cardiac output eventually results in death. My body was not getting enough oxygen or nutrients and nothing was telling my heart to get up and go. Blaine listed my problems: anginal-like chest pain, orthostatic hypotension (abnormally low blood pressure), and bradycardia (a slow heart rate). But the root causes were still unclear.

How odd that we walk around with these bodies, live in them, die in them, make love with them, yet know almost nothing of their intimate workings, the judicious balancing act of homeostasis, the delicate architecture of their organs and systems, or the varying weathers of their private, internal environments. Up to this point my living and breathing had been an act of faith. I existed but I didn't know how. I was a stranger to the body whose consciousness said, "I know myself," which meant only that I had decoded the brain's electrochemical message that told me to think such a thought.

I lifted the bedsheet. All I saw as I looked down was a pale container, skin whose bruises and cuts were only ornament, ruby and onyx jewels. How could I have been so uncurious? If I held a match to my heart, would I be able to see its workings, would I know my body the way I know a city, with its internal civilization of chemical messengers, electrical storms, cellular cities in which past, present, and future are contained, would I walk the thousand miles of arterial roadways, branching paths of communication, and coiled tubing for waste and nutrients, would I know where the passion to live and love comes from? It is no wonder we neglect the natural world outside ourselves when we do not have the interest to know the one within.

chapter 5

"Fire is not following you, you are following fire," Takashi Masaki, a Japanese farmer-monk admonished me. Was this ocean into which I had fallen a blue flame? For six months actual fires had been breaking out all around me. A plane caught on fire on the runway in Denver and when we leapt to the ground from the stairs and began running—as per instructions—upwind from the plane, the fire engines almost ran over us. A month later the Dallas hotel I was forced to stay in because of another delayed plane burst into flames as I entered the lobby. In June several IRA bomb scares in the London tube sent me dashing up steep stairs to the street. And in July the spruce forest that flanked the runway at Fairbanks erupted as my plane landed and for two weeks the skies of interior Alaska were smoke-gray.

"How does it feel to have fire enter you?" another friend asked after hearing of my accident. I had no answer. Nothing of the incident remained in my retrievable memory. All the cultural references I knew showed gods throwing lightning bolts, not ingesting them, but like the young fire-eaters I'd seen as a child, stationed on the sidewalks of the Paseo de la Reforma in Mexico City, I had swallowed fire.

Lightning is a massive electrical discharge occurring in the atmosphere of the earth, as well as on several planets, and can extend from five to ten kilometers in length. There are about 1,800 thunderstorms in progress over the earth every moment and lightning hits the planet one hundred times each second. In the continental United States alone, there are forty million cloud-to-ground strikes each year.

The life of a 20,000-foot-high cumulonimbus cloud is about twelve hours. It is a city of turrets and towers made out of polygonal convection cells whose interiors are all warm air rushing upward and whose skins are moving walls of cold air. The cumulonimbus is all motion, made of raw energy and mist.

The earth radiates solar energy, warming air, which rises and expands as atmospheric pressure decreases. Taking on altitude, it gives up heat for height. The water inside the cloud condenses and droplets hang on particles of dust. That is how a thundercloud is made. It is a community of cells organized into weather factories in which rain, hail, and snow are dropped to earth, and in which lightning occurs.

Heat and cold, water and dust, that's all it is at inception, but trouble brews. A thundercloud grows unruly, as all cities do, when the shearing stresses between ascending and descending air—as with the wealthy and the poor—result in turbulence. In addition, when dense dry air from outside the cloud is displaced by the updraft, it mixes with saturated air, thus providing a constant supply of recently warmed air full of moisture. This is fed to the upwardly mobile tower. Once begun, the cloud builds on itself, sometimes rising 40,000 feet in the air.

All summer these stately empires sail above Wyoming

mountains, processions of cool heads, but inside they are dynamic, chaotic districts drawn into existence by jets of buoyant air, growing in volume and height until they bump into the upper reaches of the stratosphere. Even then, they sometimes continue upward, their turrets penetrating stable layers of air until they can go no further, then they fall back on themselves.

This is only the beginning of the violence a thunderstorm accrues. Benjamin Franklin brought lightning down out of the sky with a kite, a string, and a key. His kite was the object the cloud's electrical charge so desperately sought. Thunderclouds are Hegelian: Electrons and protons are charged particles surrounded by an electrical field that attracts charges of the opposite sign.

Inside the misty cloud-bubble, collisions occur. According to Earle Williams, an MIT geophysicist who gives seminars on lightning, a storm's convection carries water into the cold cloud-top, where it freezes into graupel. Moved vertically, these icy bits collide with other forms of moisture and the friction generates both negative and positive charges.

A thundercloud grows in a state of imbalance. Polarities change back and forth within the cloud as well as on the ground, where the earth's negative charge flips to one that is positive as the storm approaches. Soon, everything is humming with electricity, even individual raindrops. The chaotic acceleration of charge separation taking place divides the cloud into opposing territories that end in a tripole form, a plus-minus-plus structure, like a villanelle, with a strong positive at the top and bottom and, in between, a pancake-shaped region of negative charge.

For a short time the insulating capacity of air prevents the two attracted charges from meeting, but any upright object in the force field becomes a finger that reaches up, straining to touch the other. So loaded is the cloud with electricity, that its negative charge tears a path through the air, stripping off electrons and leaving in its wake positive ions. In the newly carved channel, a sudden flow of electricity slurries down. All this has taken a fraction of a second.

But lightning is not a one-way street. As soon as its tip nears the favored object or area, upward-moving discharges, called "return leader strokes," fly up to an ionized path traveling a third or more times the speed of light. When the return stroke has ceased to flow, another dart leader may drop down, in turn initiating a second return stroke, and so on. Contrary to popular belief, lightning loves to strike the same place twice, since it always follows the path of least resistance. What could be handier than reusing this ionized channel?

The Navajo word for thunder is *I'ni,* meaning "that which moans indefinitely." When the lightning stroke goes back up from the ground, the current surges to 100,000 amperes, or 100 million watts per meter of channel, and the temperature rises to 30,000°K. This heated air causes gases to expand in the discharge channel and a shock wave is sent traveling, quickly decaying into an accoustical wave whose signal, or "signature," is what we call thunder.

Thunder is nature's unique percussion symphony. Pliny wrote in A.D. 77 that it was unsafe to speak of certain kinds of thunder, or even listen to it, lest it bring bad luck. On the other hand, he also noted that Romans used thunder as a tool of divination to predict events, and listened to it to hear secret messages.

We now know that thunder's wavelength is determined by the length, duration, and total energy of the lightning stroke. Instead of reading thunder to know our fortunes, we use it to "read" lightning. For example, the lower the pitch of the thunder, the more powerful the lightning strike has been.

The violence and energy produced by a thunderstorm starts as a spark, then many sparks conjoined and flowing as if seeking to illuminate a dark patch of ground or make incandescent every window of a darkened city with its sudden, ephemeral light. The electricity inside a cloud sweeps back and forth, up and down, always seeking the path of least resistance, while the dynamics of convection works like a heart, pumping air and moisture up through the valve of the cloud and pushing electricity down through an artery. Nothing is ghostlier than lightning's light: pale, colorless, it serves up frozen instants, then disassembles those phantasms and delivers them back to darkness.

We swim in an ocean of air, in magnetospheric, ionospheric, and tropospheric currents bound together by a global electrical circuit. The surface of the earth gives off a negative charge, which is met equally by a positive atmospheric charge whose conductivity increases with altitude. Galactic cosmic rays bring positive and negative ions into the earth's atmosphere but it is thunderstorms that generate huge amounts of electricity. They are the factories that keep the global circuits going.

There are intracloud, intercloud, and cloud-to-air flashes

—heat and sheet lightning, discharges that never touch the earth. Rocket lightning sends horizontal sprays of light across the tops of windblown clouds, and ribbon lightning's stroke is one that has been separated in the channel by wind, thus giving off a double image; cloud-to-ground lightning (the kind that struck me) breaks into luminous fragments, like a necklace of pearls, but no name has been given to the wild lightning that zigzags in all directions at once.

Ball lightning is controversial, often being discounted the way UFO sightings are, because it is not scientifically understood. Are these luminous globes a brew of storm chemicals (oxygen, hydrogen, and nitrogen) or a "brush discharge"—atmospheric electricity bound together somehow? Or is it a collision of charged dust particles, raindrops, and ions? Sightings of these glowing spheres have been recorded continually from ancient times, as crazy, rolling emanations kicked down from the heavens or from some Jovian beach, sometimes bouncing, sometimes rolling down chimneys or slipping into open windows and disappearing under beds like fiery dust motes. They can move fast or slow, hesitating before they roll, and they can either move against a wind, in defiance of physics, or go with the flow. Not rolling stones, they are bodiless, centerless, with no hard nuclei around which sparks can spin, yet they hiss and sparkle and appear, interchangeably, as balls of gold, blue, white, green, or red. Intensely bright to the human eye, they are often "cold," or else hot, as in the case of Diane of France, who on her wedding night, in 1557, saw a ball of light pass around her bedroom in an erratic course, finally bouncing onto her covers and burning her clothes and hair.

A "fire dragon" is what Gregory of Tours called it. But as he had already seen flames emanate from certain sacred relics, he was not surprised when a ball of lightning rolled over the top of a religious procession he was leading; he simply proclaimed it another miracle.

" 'Look aloft,' cried Starbuck. 'The corposants! The corposants!' All the yardarms were tipped with a pallid fire; and touched at each tripointed lightning rod end with those tapering white flames, each of the three tall masts was silently burning in that sulphurous air, like three gigantic wax tapers before the altar."

That's how Melville described Saint Elmo's fire, also known as "corposants," from the Italian *corpo santo*, meaning "holy body"—those sudden fires that appear on yardarms, at the tips of wooden masts, between the horns of cattle, at the tips of airplane wings, and around metal objects, especially at high altitudes.

Saint Elmo's fire was named for a fourth-century Italian bishop who was rescued from drowning by a sailor and ever afterward swore to give a warning of approaching storms to sailors at sea. But in fact, by the time Saint Elmo's fire appears, one is in the midst of a storm—too late to do anything about it.

In the journal of his second voyage, Columbus noted: "On Saturday, at night, the body of St. Elmo was seen, with seven lighted candles in the round top and there followed mighty rain and frightful thunder. I mean the lights were seen which the seamen affirm to be the body of St. Elmo, and they sang litanies and prayers to him, looking upon it as the most certain that in these storms when he appears, there can be no danger."

This "holy body" is actually the swarmlike glow of atmospheric electricity, too low a discharge to be harmful, but high enough to light up the highest points around. Sometimes masses of flying insects carry this incandescent charge and are mistaken for UFOs.

Transient luminous phenomena, like the transient coronary spasms I now experience—little cramps in the heart muscle—have been observed by unmanned spacecraft in the atmospheres of Jupiter, Venus, Saturn, and Uranus, whose cloud layers, made of ammonia ice and filled with particles of water, are spawning grounds for lightning. Do they have afternoon thunderstorms on Jupiter? Does Jupiter have afternoons at all?

Storm-related electrical discharges are shrouded in mystery. No one theory of cloud electrification can account for the prodigious amount of current produced in a thunderstorm. Being in the right place at the right time to get samples—as I was—is not an experiment for which you volunteer.

chapter 6

Thirty years ago, my sister, Gale (so named because a gale hit Boston Harbor the night she was born), some friends, and I stole a boat in the middle of the night and sailed it out of the Santa Barbara harbor. Suddenly we were becalmed and the current began pushing us toward the breakwall. We could hear the foghorn drone and waves crashing against a rock. With no running lights and no power, we were dead in the water. Out of that darkness a steel hull appeared: it was the local Coast Guard cutter. My father, stern-faced and displeased, stood in the bow.

Now I'm here because of another rescue by my parents. I was born in this hospital, rescued from formlessness, given a body. Am I to die here too? These are the shores between which we are all suspended: the perilous breakwater and the safe harbor, the foggy night and the bright day.

Just before dawn on the second day, a man on a ventilator, who had not moved since I'd been admitted, died. The alarm went off, crash cart swept by, doctors and nurses hovered over him trying to restart his heart. He lay naked, rotund, a blue tube going down his throat as they worked on him, but too much of his heart muscle had been damaged: it could no longer pump blood. After, the unit was quiet. One of the nurses who had been working all night sat on my bed and wept.

Death always feels like a failure. I told her about how, on the ranch, animals died at dawn, as if to say the energy of facing another day was too great a burden, and how the dead piles outside the lambing shed reached the roof by evening, but the ones we saved made it worthwhile.

Another alarm sounded. Blaine was on the floor, and I could hear him giving directions, cool-headed and firm, and saw his expression of relief as the woman's heart started again.

In a few minutes he appeared at my bedside. With his eyes closed in concentration, he leaned over to listen to my heart. Beads of perspiration jeweled the deep furrow in his forehead and his damp hands trembled slightly.

"What a way to start a morning," I said, but he only smiled.

"Are you always this cheerful?" I inquired. I wondered if there were any dark corners in this man.

"No, I just hate facing pain. When I'm at the dentist I always ask for double Novocain."

When he finished the exam, he sat down in a chair. He had a sensual handsomeness and a farmer's big hands and feet.

"My friends always teased me at medical school about being too happy. I made it a rule never to study on weekends. What good does it do to worry?" I asked if he got "As" anyway, and he said, "Oh, I guess so."

Med school was Cornell with a residency at New York Hospital, a stint at an army hospital in Albuquerque during the Vietnam War, a rotation on the Navajo reservation, and a fellowship in cardiology with Norman Shumway at Stanford University, where Blaine helped perform the first heart transplant.

"It was a very exciting time. In the sixties, cardiology was really taking off. We were learning so much, and the technology was improving so rapidly. Every week there was something new. We were suddenly able to save lives that could not have been saved a few months earlier."

Listening, I sat on the edge of the bed, dangling my feet in the foggy air flowing in through the window.

"I'm in medicine as a practitioner—not a researcher—because I love people. I'd always wanted to be a doctor, but my parents didn't have the resources to send me through med school. I thought of enrolling in engineering school instead, which would have been a disaster. Everything I built would have fallen down. At the last moment I was offered a generous scholarship to Swarthmore College that allowed me to use my savings for medical school, and I never looked back."

He stood to go, then told me he had already read one of my books and had begun another. "I'll never get any reading done if I sit here all day," he said, grinning. "Catch you later."

Later a nurse came in and encouraged me to get up and walk around (good for low blood pressure). She said, "Just tell us where you're going so we can keep track of you." One loop through the unit was all I could manage before retreating to bed.

During the night I had another "sinking spell," as my mother called them. Even lying flat I could not stay conscious. When my heart rate fell to thirty, the buzzer went off and nurses appeared.

But the center did not hold. Each room was a composite flower's petal exploded out, propelled by fire. . . . I was dying. Hummingbirds circled my head, separating oxygen from blood with their beaks. I gulped the rich dessert of air. Sandhill cranes flew through the room, way up near the ceiling, their cries growing fainter. I was going the other way. . . . Then I heard a nurse say to me: "Don't worry, we won't let you die."

How long have I been wandering? It was night when I started remembering again and I had been skating. In some places the ice was black, in others checkerboarded translucent and white. But the blades would not grip. I tried to glide but my knees faltered. Then I saw down into the ice: miles and miles of transparency.

It was the dogs, not the nurses, who brought me back. They had done this so many times, the images of their rescue had become abbreviated: their sleek coats, both dark and blond, sideswiping my face; the sudden tug of the harness; then, bright air. The hospital room was quiet and I was inhaling oxygen. Between darkness and light, death and life, one breath and the next, there are these gaps. Even the lightning stroke is an opening, a vein in the heart of a storm, a passageway carved in the ivory of cloud. The open window behind my bed let mist in the room. I could hear the foghorn's warning drone: BEWARE!

In the Bardo Thödol, known as the Tibetan Book of the

Dead, *bar* means "between"; and *do*, "a landmark that stands between two things"; joined together, *bardo* means "gap." It refers to that wandering state between life and death, confusion and enlightenment, neurosis and sanity. The past has just occurred, and the future has not yet happened. In the bardo of the human realm we experience the body as illusory. Our relationship to our own existence and nonexistence is lukewarm. The whole world is a hiatus; the gap is not just a widening in the road before the next bend, it is where the road falls off the cliff.

The bardo has also been described as a vast and desolate plain littered with corpses and bleached bones and feral animals feeding on remnant flesh, a plain that is crowded and empty at the same time where animals copulate wildly, fall away from each other, and move on. Then it's a gray ocean again with no surface and no bottom, no reference points, no lighthouse, no breakwater guarding the harbor, no guiding light to lead me home. The bardo state occurs not only at the moment of death or the moment before death, but all during our lives; the bardo is the uncertainty and groundlessness we often feel.

Chogyam Trungpa Rinpoche, who made the first translation of the Tibetan Book of the Dead for English-speaking Westerners, was born in a cattle shed to very poor parents. Trungpa was "discovered" as the incarnation of the tenth Trungpa Tulku when he was thirteen months old and taken to the Surmang monastery, where he was installed as abbot. Years later, when the Chinese Communists were invading Tibet and pursuing the tulkus, such as the Dalai Lama, and Trungpa and his entourage rode and walked across Tibet in the hope of reaching India. They had no maps—none existed. They began the dangerous crossing of the Himalayas in mid-winter, on December 15, 1959, under a full moon, and when they ran out of food, they boiled saddle leathers to make broth, since they were not allowed to take the life of any sentient being. Many of the great monasteries and most of the ancient Tibetan culture was being destroyed behind him. He would never see Tibet again. Yet he referred to his journey as "his trip to freedom." In fact it was exile.

Trungpa said the Book of the Dead is not only an instruction on how to die or help the dying but should rightly be called the Book of Space, the Book of Life, or the Book of Liberation—a primer in the physical aspects of death that arise as soon as we are born, as well as psychological styles of living and dying. A greater familiarity with death emboldens our love of life. The bewilderment and desolation we experience in the bardo—"I wander in the bardo state alone . . ." one of the prayers says—translates into a sense of delight with the play of opposites, the way solidity and attachment breaks down, with what Trungpa called "the indestructibility of impermanence."

How far have I wandered? At dawn Blaine burst into the room, yanked the oxygen tube out of my nose, and smiled. "You don't need that now." He sat on the bed and told me about four migrant workers in the Salinas Valley who were hit by lightning the day before—but he couldn't find out anything about their condition. They had been released, as I had been, with no follow-up or monitoring. Blaine's visit pulled me back into life, into the world of newspapers and human problems; his curiosity and enthusiasm made me want to be alive. As he was leaving he told me he had called in another cardiologist, whose specialty is cardiac electrophysiology, to help figure out what was wrong with me. "We're treating the results of your problems, but we don't know the root causes. We don't know very much about how electricity affects the brain and heart." An unpleasant test was scheduled for the next day.

Suzanne, the nurse, checked the pulse in my ankles, groin, neck, wrists. "If you go outside to sit this morning, don't mind the underwear," she said, pointing to some clothes draped over a chair on the patio. "A homeless person came into emergency and he was dirty, so I washed his underwear for him before I went on duty. He had been walking for days, he told me. Just walking."

I couldn't believe I was in a physical state that prevented me from walking. Walking had long been my liberation. Every summer on pack trips I tied my saddle horse into my husband's pack string and walked fifty or sixty miles through the mountains at ten thousand feet. And alone at the ranch while my husband was in the high country in another part of the state, I often ran crouching, over hundreds of acres. Now, to think of walking was absurd. I could barely lift an arm; to think of not walking was even more absurd.

All my senses, perception, energies were telescoped down into that tiny room: a "hospitalscape" with no horizon. Each time a nurse entered, I could sense immediately his or her state of mind. She didn't need to speak, her presence told all: how preoccupied she was with her own concerns, the degree to which she could give herself over to any patient, how deeply she lived in the moment, her nose in death and her elbows in the messiness of living.

Trungpa's Book of Space is also a book about charnel ground, which is the full aspect of emptiness and desolation. This hospital unit for the acute care of cardiac patients was a piece of the Tibetan Book of the Dead.

Once the bewilderment is gone, the Book of the Dead tells us, the gap is an expanding universe in which space is sharply defined, where there is "radiance with no radiator, accommodation with no territory, fulfillment with no motivation, communication with no receiver."

At sunset I made my pathetic loop around the unit, then ventured out onto the patio. The entire western sky was red. "It's from the ash," the nurse who came to check on me said. Today's volcano is Mount Pinatubo, which erupted in the Philippines in June 1991, and whose ash still circles the globe. Lightning often occurs close to volcanic eruptions, and "flashing arcs," fly out of craters and spread in all directions with the velocity of sound.

I sat down in front of the wall of flame. The homeless man's long underwear flapped in the breeze. He was somewhere in this hospital, lying in a bed, no longer walking. Nor was I. Lightning had entered me twice and now I was a burnt shell with nothing in me that could attract fire. The curtain of red was unmoving. Then darkness came down.

I returned to my nest. As fog moved into the room, the illusion of place disintegrated. Already the end of the bed was gone, my legs, the nurses' station. The heart monitor's red numbers flashed backwards on glass doors, as if I had been turned inside out, life turned into death like a piece of suede, smooth inner skin exposed.

When you are sick, days lie together haphazardly, like empty containers: each one counts because it means you are still alive, but the details of the greater world are unclear. Sickness entails a hiatus, a gap in habitual activity, an interval during which one is suspended motionless.

From my all-interior-view room I could read the names of patients handwritten in bright colors on the chalkboard behind the nurses' station. Every day the list changed—people either died or were discharged. My name was still there but it meant nothing. There was no longer a narrative to follow. Even the character of death had dropped out of my tale.

chapter 7

Joseph Ilvento, the young cardiac electrophysiologist called in by Dr. Braniff, specializes in the effects of electrical impulses on the heart. He spends his days sleuthing the electrical clues that might explain why a heart has gone into arrhythmia, tachycardia (rapid heart rate), or bradycardia (which is what I have, a slowed-down heartbeat), as well as "conduction disturbances" (electrical messages that have misfired). Much like an astronomer who can never see with the naked eye the galactic events he has devoted his life to studying, Ilvento deals with the consequences of electrical events—those secret journeys of impulses from synapses, down nerve fibers, through ganglia into the heart—which he can never touch with his fingers or see.

Cardiac muscle is absolutely unique. Unlike skeletal or smooth muscle, the cells that make up this tissue have as their main purpose regular, periodic constriction. They are designed to drive the pump, to make the heart contract and release, not as a response to a voluntary impulse but spontaneously. Even the design of the fibers along which cells are clustered is different. While skeletal muscle fibers are long and stretchy and can contract at will to variable lengths, compliments of the central nervous system, cardiac muscle consists of crosshatched, branching filaments whose interconnections place cells so closely together that electrical impulses are conducted with terrific speed and accuracy.

Myocardial cells have bulbous heads that, once activated, move in oarlike motions, rowing and rowing, bending the filaments in the direction of contraction, then disconnecting them as each rowing motion comes full circle, only to begin again. Under a microscope, normal cardiac muscles are dendritic and branching, and their fibers are arranged in patterns that look like lightning. When cardiac arrest is reversed, the tiny myocardial cells row through the frozen sea of tissue, pushing the pump into action again.

I found Dr. Ilvento rather serious, but, according to Blaine, always the optimist, Ilvento is likable, "because he makes great pizza." But pizza, which I rarely eat, was farthest from my mind when the head-up tilt-table test was described to me. "We have to find out why you're passing out all the time, why your heart rate suddenly plummets. What we'll be evaluating is your 'unexplained syncope.' " "It's going to be unpleasant," Ilvento said, though he might have said horrific.

I was strapped down—wrists, waist, ankles—to a table. Dave Wallace, a nurse assigned to Ilvento's patients, velcroed blood-pressure cuffs and attached leads to an EKG machine. I asked if they were going to torture me, told them that even if they did I wouldn't talk. Dave smiled. "It's not exactly torture . . ."

The idea was this: I was to lie flat on the table for twenty or thirty minutes while blood pressure and heart rate were continuously recorded, then the table would be tilted straight up and I would hang, suspended in the standing position, for as long as it took—up to sixty minutes—to see if I lost consciousness.

That morning I learned the meaning of panic—just in case I had missed that experience in my life. Before the position of the table was changed, Dave, the nurse, jiggled the straps. "Don't worry, you won't fall. Just relax." Then they tilted the table straight up. It's normal for blood pressure to drop eight or ten points when a person moves from a sitting to a standing position but mine slid. I had been crucified in my coming-back-to-life dream; now I was not dead in gray water but hung ridiculously to a table in a bright room surrounded by people who were trying to keep me alive by examining the mechanisms that were making me die.

Drooping from my gallows, I was asked repeatedly by Dave and Ilvento how I felt. But it was the same old story: the room was black. Clamminess turned to a drenching sweat; my breathing came fast and the terrible, elephantine heaviness invaded my body again. The nurses were vague forms as they checked the monitors: heart rate and blood pressure dove.

It's not true that I wouldn't talk if tortured. As the black hood dropped over my thinking, I began to plead for a reprieve: "Please put me down, please . . ." The instinct to lie flat was indomitable. It's the life-saving desire of the body to get more blood to the head. Quickly it became apparent to Ilvento that the "conductor disturbances" occurring in my body had to do with "vasovagal syncope"—the failure of the blood vessels to constrict. Blood and oxygen had drained into my feet and legs, and though I begged repeatedly and vehemently to be laid flat, they refused, simply watching my body go limp. I tried hard to hold out, to hang on to light and form and discursive thought, but the balloon burst and I was gone.

What is the architecture of a blank, and how long can it last? What happens to memory when one becomes unconscious, does the synaptic gap go dark? Do the neurotransmitters drift aimlessly?

Ahead of me it was dark but I could see the dogs' glistening fur. We were sledding in moonlight. A needle was slipped into my vein. Atropine. I slid from one black puddle to another on a road made of limitless gravity. The needle in the vein was like a needle in a haystack, a needle flying between galaxies. Blackness prevailed. The dogs who rescued me had dark fur. I foundered in a dream about horses whose back legs—hips, hocks, cannon bones, pasterns, hooves—were cut up and laid at odd angles in a wheelbarrow, to be carted away. The dogsled carried me back into the day.

I felt Dave's hand on mine. He was saying something to me but I couldn't answer. Ilvento was on the phone reporting to Blaine:

"We got a positive result. She was out in fourteen minutes so this answers our questions. I think we can safely rule out most everything else. It's vasovagal."

My face was gray and the back of my hospital gown dripped sweat. Almost unbelievably, they repeated the test after I had rested. "We won't let you go all the way out this time," Ilvento promised, but I was skeptical. I felt like telling him I hated pizza. They tilted me up and quickly I felt faint. The black tunnel tightened around me—a noose—as if I had been sent to the gallows. Or was I being crucified again?

When they laid me down, atropine was started. It wasn't that I had suffered any pain, it was only discomforting. Later, dazed and unsteady, the nurses led me back to bed, like torturors leading the tortured.

You don't have to experience everything life can throw at you—torture or ecstasy—to fuel empathy; just a taste is enough. I vowed to send money to Amnesty International for victims of torture. When Blaine came by that afternoon his jauntiness rubbed me the wrong way for a moment. He sat on the bed and looked at me. My face was gray. It wasn't necessary to ask how I was feeling, but he did anyway.

"Everything is still dark," I said. "I feel washed out . . . like I'm swimming and there's nothing to hang on to."

"Here," he said and held both my hands in his.

At the end of rounds he returned to my room. I was slightly more bright-eyed. He sat down and explained the findings of the tilt-table test. "You've suffered an electrical insult to your brain stem and have lost the ability to vasoconstrict. You have no vascular tone in the smooth muscles of your blood vessels. You've also lost the ability to increase your heart rate. The sympathetic and parasympathetic nervous systems usually counterbalance each other. When the heart and blood pressure get too high, a message goes through the parasympathetic nerves to the brain, which secretes an inhibitory chemical that slows the heart down. The sympathetic system does the opposite: it speeds things up. I think your sympathetic system was burned by the lightning, allowing the parasympathetic vagus nerve to take over and have its way. It keeps telling your heart to slow down, because no excitatory chemicals are being released."

I stared blankly. "Could you repeat all that sometime when I'm feeling better?"

"Don't you trust me?" he said, laughing, then put his hand on my forehead and told me to rest.

In the morning I noticed Blaine's eyes had changed color: they had been brown and green and now they were gray and blue. My mind was mush. A tight band around my chest tightened and eased, depending on the threadiness of my blood pressure. Sometimes during rounds Blaine stopped to talk, and other times I heard him talking to nurses. My own hazy monologues continued: I wondered what the interior geography of my body looked like or if my insides could be read like illuminated manuscript.

Between visits I wandered. The peregrinations felt intergalactic. Fingering the color fold-outs of Gray's Anatomy, I traced nervous systems, blood vessels, intestinal coils, musculature, spinal cords, and the convolutions of the brain. A body is a separate continent, a whole ecosystem, a secret spinning planet. The brain looks Vesuvian with its breaks and draws called "gyri" and "sulci," its fissures and fjords. The large internal fold of the cortex is called an "insula," and near the cerebral aqueduct dividing the midbrain grow stalks of neurons called "infundibulum." There is a "vermis," and "arbor vitae," a "pons," and the small swellings of the medulla are called "olives." Who named these parts of the body? I asked Blaine. He didn't know. I thought it must be a walker like Thoreau, or a mad geographer, an Arctic explorer, or someone making a miniature garden of the brain.

"Look." I showed him a picture. How the brain sits on its spindle, like a globe, the nodding head tilting on its axis, how the nervous system is a series of branches sprouting from that *axis mundi*, how each thought passing through is a separate ecosystem.

"The brain has a hundred million nerve cells and is seventy-eight percent water," he replied.

More fog rolled through the window like reason, trying to hide things, smooth things over. The room was a skullcap. When Blaine returned in the afternoon, I asked, "How does it all work?"

"What?"

"The systems of the body."

"That would take years," he said.

"I don't have years."

He gave me a surprised look.

"Yes, you do," he said.

"Tell me anyway."

He thumbed backwards to a picture of the whole body. No one part functions as a separate entity: the brain does not coolly dispatch messages as a computer does, nor is the nervous system just a system of highways, he explained. It is the communicator inside us, at once both pathway and messenger, though, in my case, the pathway needed repaving and the messenger was dead.

I smiled and he continued. The nervous system is composed of a network of cells that extends throughout the body, receiving information about the internal and external environments, assessing that information and then sending signals to organs that cause an appropriate response. It is a great branching tree dividing down into separate systems: the central and peripheral, of which the autonomic nervous system is a part. My autonomic nervous system had been damaged by lightning. Regulated by centers in the cerebral cortex, hypothalamus, and brain stem, it is divided further into two subdivisions that must constantly balance each other—the sympathetic nervous system, whose purpose is to stimulate activity in the heart, blood vessels, stomach, and sweat glands by releasing an excitatory chemical called norepinephrine, and the parasympathetic nervous system, which does the opposite, releasing an inhibitory chemical called acetylcholine, which slows things down.

"Because your sympathetic nervous system got fried by the

lightning, your vagus nerve has gone wild. There is nothing to tell it to stop sending its messages to slow everything down," Blaine said. "It's a big long nerve that innervates the lungs and heart, liver, stomach, pancreas, small intestine, and kidney. That's why it's called vagus. It wanders around in the body and tells your heart to slow down and your blood vessels to stop contracting until you have no blood pressure at all."

I have a wanderer inside me, a vagus nerve, a vagrant that does whatever it wants. Any time the fine balance of the body's internal civilization is upset, the whole intricate communication breaks down: when the heart slows or stops, cells, synapses, and receptors become confused and leaderless. Where is that oxygen, they wonder, where are the nutrients that keep us alive? Homeostatic panic ensues. For a while the self-regulating universe of the body tries everything it can to compensate, but if that doesn't work, then it's good to have a doctor nearby.

The next day I asked Blaine and Dr. Ilvento to explain what actually happened to brain and nervous system cells when a great deal of electricity passed through the body. Together, they speculated: "The mylenated sheaths—the fatty white matter that protects neurons—may have bubbled and melted; cells died and were sloughed off [there are cells in the body that do nothing but cart the dead ones away]; neurotransmitters with their excitatory chemicals dried up. The path taken by lightning inside the body became a desert."

I touched the place on my back where I had felt a burn after being struck. Blaine touched it. "Does it still hurt?" he asked. "It's sore," I told him.

"This is your exit wound. It's near the heart, right where the sympathetic nerves hook in," he informed me.

Blaine ordered new medicines and larger dosages. Norpace made my heart beat regularly no matter what messages it received from the brain, as well as working on the vagus nerve by blocking acetylcholine. The Florinef helped my body retain water. It was a simple gardener's idea: watering a plant maintains the turgidity of a plant's stem—in the human body, retention of water helps raise blood pressure. "Eat lots of salt and drink strong coffee," Blaine said, laughing. "Can you believe a cardiologist is recommending these things? Also, stay cool, no Jacuzzis, or hot baths, no alcohol, and keep your feet up."

Two days later I was released. My father came for me. Leaning on his thin arm, I walked out of the cardiac care unit barely able to comprehend what was around me: waiting rooms full of strangers, miles of polished floors, then cars, sidewalks, palm trees and sycamores, and sea smell. It was all I could do to put one foot in front of the other.

During the drive home the sides of the world heaved up and folded in on me, and the road was a concavity dropping out from under the car. With my seventy-five-year-old parents, I was still their child, dependent on them for shelter, food, rescue, and survival. Even their pace was too fast for me. "Now you know," Blaine said, "why patients are called patients."

chapter 8

I was once the young girl who wanted everything to be fast: horses, cars, wild and turbulent rides in my father's small plane. My appetite for life was clamorous inside. Now even that din had been quelled. I doddered around the house, lay on the floor with my feet up in the air to get more blood into my head, and from time to time begged my mother for oxygen.

Life consisted entirely of rest, round-the-clock pill taking, TV watching. Norpace blurred my vision, so reading was difficult, nor did I have the concentration for it. Someone else's story was too big to swallow. All my wits and brain power went into maintaining my own life: I was the homeostatic engine trying to reestablish equilibrium in a body seriously off-kilter, which is another way of saying I stayed close to couch and bed. If I had any ideas about going out that first week, waves of low blood pressure knocked me back down.

On my next checkup at Blaine's office, I informed him that I was going to London in two weeks for the opening night of the ballet for which I had been writing a text all summer. Undaunted, he smiled, but I didn't believe his casual air. "I'm not going to miss it," I repeated, just so he knew I was serious, though down deep I couldn't imagine how I would get myself there.

During the second week I made a supreme effort to face the world. After all, if I was going to fly to London by myself I had to raise my energy level. On a good day my mother drove me to the beauty shop, whose atmosphere of care and silly gaiety cheered me. My singed hair was brittle as wire. A cheerful young woman cut and drenched it in conditioners to bring it back to life. Staying conscious during those visits was tricky enough, but I managed. Other days we went to lunch, always taking oxygen along. I had a craving for hamburgers, though a few bites filled me completely. It was only tastes of life that I wanted; biting in and chewing was still too much for me.

One morning I visited a Chinese acupuncturist. He was horrified at the medicines I was taking. On the table, with needles sticking in my head and fingers, I nearly passed out. He immediately changed the placement of the needles and I came around quickly. He showed me a pressure point on my little finger to use if I started to black out while traveling.

At the end of the week I lied to Blaine. Two weeks had elapsed and I was to fly to London the next day. Driving for the first time since getting out of the hospital, I had to pull off the freeway three times to keep from blacking out and lie down on the front seat with my feet pushed through the window. Figuring there was nothing to lose, I tried the acupressure point Dr. Wang had showed me and finally revived.

In the parking lot behind Blaine's office, I jumped up and down before going in. He took my pressure and sure enough it passed muster, though it was by no means normal. A little later, when he asked how I was feeling, I said, "Great." My determination to get to London would get me there, I presumed, though I sensed Death had not forsaken me altogether. I knew how easy it was to die, and, by God, I was going to pack a few more things in before Death took me away.

chapter 9

My travel agent made sure I was booked on uncrowded flights, so I could lie down, three seats across. It was my only hope to keep from passing out. I wrapped my legs in Ace bandages to prevent blood from pooling in my feet and licked palmfuls of salt, chased with water. The trip to London was broken by an overnight stop in New York. My publisher and his wife picked me up at the airport and I spent the night in their Brooklyn home.

In the morning, a driver took me to the airport and I had to lie flat in the back seat to stay conscious. The plane was like a drug that lifted me out of the human world with its sorrows and entanglements. I stretched out with pillows and blankets. In the past I had preferred sleeping on hard ground, gravel walkways, granite ledges in high mountains. Now the simple comfort of a soft seat was an unexpected gift. To surrender and sink and still be carried forward all night across the sea was trancelike, as if I had stumbled upon life beyond death that magically becomes life again. The Atlantic was a great river running north and south under me, a channel through which lightning sometimes pulsed, but I was shooting across it at right angles, far above, where lightning could not touch me.

London cheered me as nothing else had for a long time. During the summer months of June and July, I had collaborated with a British choreographer, Siobhan Davies, on an evening-length ballet and had only gone home to the ranch in early August, to rest up, ride my colts, and help neighbors move cattle before returning to work with the company again—and that was when I was struck by lightning. Now opening night at Queen Elizabeth Hall in the South Bank Theatre complex was a day away. The dance, the music, and my text had been put together without me.

I felt safe staying with my friends, Siobhan ("Sue") Davies and David Buckland—who designs her sets—because he had a heart problem too. When an infection from an abcessed tooth traveled to his heart, David suffered from endocarditis—a fast-moving infection that caused nine cardiac arrests. He is now alive, thanks to a pacemaker and a mechanical mitral valve, whose ticking could be heard in the quiet theater when I stood next to him. His cardiologist lived nearby, and I knew David would understand when and if I needed help and how urgent that need can be.

The next twenty-four hours were spent in the theater. I lounged in velvet-covered seats, legs propped up, amazed at the spectacle of our dance. All summer we had rehearsed in a crowded and unglamorous hall in a Soho synagogue. Now, for the first time, I saw the spectacle of our creation.

The dance—choreographed from a poem cycle I wrote after being in the Canadian high Arctic—was both stark and sensual, fluid and athletic, graceful and erotic, with impossible contortions of male and female bodies that resolved into gestures of solitude or ardent embraces. Above them a single revolving light stood for the Arctic's circling sun. The stage was blue: a blue backdrop with David's sixty-foot-long painting of Sue's body floating over the stage and a blue ground cloth nailed tight to the stage—the floor of the Arctic—from which dancers rose and into whose shadows they walked.

The day of the performance was also the day of the company's one and only technical rehearsal at Queen Elizabeth Hall. The two sixty-foot painted cloths were stretched and fastened, lighting cues were dictated, music was heard for the first time, and dancers struggled to get their timing just right. Peter Mumford, the lighting designer, had just flown in from St. Petersburg, where he had been working with the Kirov Ballet. When he saw the stage during the first run-through he decided the lights were all wrong and with five hours to go before the performance began he redesigned every cue.

The ground cloth tore the dancers' feet, and when they bled the canvas had to be retouched to cover the stains. There were blips and extra words no one had heard before in the music-and-text tape. I had recorded my poems in a studio and these had been cut into the electronic score. While the composer recut parts of the tape, we fiddled with sound levels, and it was the dancers who readjusted their timing—a process that wasn't completed until a few moments before they went onstage.

Soon enough ushers appeared holding programs, doors were opened, the stage lights dimmed. The last lighting cue had just been given moments before. Slowly the thousand-seat theater filled. I went to the lobby to greet two American friends. Outside on the terrace overlooking the Thames, the usual gray London sky had darkened with storm clouds. As the chimes rang, announcing the beginning of the performance, thunder clapped and lightning, rare in London, crackled over the river. David ran outside and grabbed me: "For God's sake, get inside!" Just as I found my seat, the lights dimmed and the dance began.

Seeing our collaborative work unfold on a large open stage—no proscenium, no rules by which we can declare what in our lives is fiction and what is "reality"—as if the two were different, as if it did not all come from the same imaginative source—was a gift, a new life, canceling out for a moment the huge and ungainly blank that lightning-induced amnesia had deposited in my private narrative of how things had been and, therefore, how things are now. The lines I had written for the dance seemed an echo: "You walk inside yourself on roads and ropes of blood vessels and tendons, you walk inside yourself and eat weather . . ."

Gladly I made for the shoreless shore of an open stage that stands for life with no boundaries, for the "bardo" of uncertainty and for whatever transpires inside the synaptic gap. I watched the line of my grappling hook play out, loose and shimmering in the dark until it came tight, fastening me to those elastic dancers, to David's blue stage of shadow and memory ticking with his pacemaker's time, and the moving planes of light that Peter had made, and together we jigged into the night.

■ ■ ■

It was time to go home. I was desperately tired. The last leg of my journey began to feel Odyssean. Overestimating my resilience, I had planned, on my return home, to go on a sixteen-city book tour for my publisher to promote a new book. Ten minutes into the first of these readings I had to excuse myself and sit down before I fell. The next day I flew across the country. Just as my plane landed in San Francisco the flames of a fire that would blacken much of Oakland broke into view and while driving to Sausalito, white ash thickened on the windshield—some of which, I found out later, was the burning pages of other writers' manuscripts.

Flying from Portland to Seattle, I experienced what felt like a heart attack: deep anginal chest pain. When I got off the plane I was gray, sweating, breathless, and my left arm felt numb. My heart was trying to pump enough blood and oxygen to raise my blood pressure but couldn't keep up with the demand. I rested for five days in a posh suite that previously had been inhabited, the maid informed me gushingly, by Jeff Bridges, a movie star. She seemed dismayed now that she'd have to bring coffee to an ordinary mortal with chest pains.

The president of my publishing company faxed this message: "Stop the tour now. Go home immediately. A dead author does us no good." But I had to rest another few days before I could travel.

I finally boarded the plane that would take me to California. It was an airbus, large and completely full. Just before liftoff there was a loud explosion. The plane lurched, then the pilots slammed on the brakes so hard, people were thrown about. My forehead hit the seat in front of me and my heart seized up—pure stress—until I had trouble breathing.

The plane came to a stop, then turned and taxied back to the terminal. No soothing words from the captain; no words from anyone. Just a stunned silence. I asked the young man next to me to press the buzzer. When the flight attendant came, I told her I thought I might be having a heart attack. She brought oxygen—though she looked as pale as I did. I popped nitroglycerin, and offered her some. She didn't know if I was making a joke. The pain stopped, then after ten or fifteen minutes started again. I breathed in and tried to relax. This was beginning to get funny, except I was too tired to laugh. If only I could get home and lie down and not have to get up again.

Hours later we were on a new plane bound for Los Angeles. The friends who met me said it was too late to drive to Santa Barbara; they had reserved a hotel room for me, and though I was frightened to stay alone, I didn't say so but went obediently to bed. Not knowing where I was when I woke each morning was no longer a surprise. The surprise was that I was alive.

chapter 10

Water puts out fire. The restoration of my health depended on it. If I stopped following fire, simply stopped, perhaps fire would stop erupting all around me. In the morning I was driven to my new abode, an airy, spare beach house rented from a friend. Though I had been born in Santa Barbara, I had never known how the city was named or who Saint Barbara was. Now the coincidence seemed uncanny.

In the 1700s, after being caught in a fierce thunderstorm, the secretary to the Portuguese explorer, Juan Cabrillo, who was passing through the channel, gave the Chumash village the name Santa Barbara in thanks to the saint for keeping him alive during the storm. He had chosen Saint Barbara for a reason. Born Barbara Dioscorus, she was the daughter of a wealthy man who became enraged when she converted to Christianity. In revenge, he had her beheaded, but immediately he was struck by lightning and killed. After, Barbara was made a saint who protected those threatened by thunder, lightning, and fire. Perhaps I had come to the right place at the right time.

In my airy living room, I lit a candle for Saint Barbara and read about amulets, household protectors, good luck charms such as the leaves of elder and bay trees, holly, ivy, and mistletoe, houseleek grown on a roof, mugwort, orpine, Saint-John's wort, or rosemary, all of which are said to protect a house from lightning.

For so many years the ranch in Wyoming had felt like the center of the world. In its wild solitude, communities of plants and animals came into existence and died; the seasons battered us; friends came and went; hard work engraved itself in our skins. Now I was someplace else and alone. No dogs greeted me, and there were no chores to be done in the morning.

Simple and spare, the beach house shook on its pilings each time a wave broke against sand. Floor-to-ceiling windows faced the ocean on one side, and mountains and a salt marsh studded with egrets and blue herons on the other. From my parents' house a mile away, one could see right through my living room as though through an X-ray in which I had been caught trespassing.

In my coming-back-to-life dream I had been crucified and suspended in the ocean. That's why, now, I wanted to live at water's edge. Here, I would surrender to whatever swam through me: death nailed to life. Here, I would be restored.

The cruciate form is an arbor vitae, a tree of life, a human with arms outstretched in surrender. The horizontal bar is the road we travel, the linear direction of a life, and the vertical stave is the spiritual elevator that lifts and drops us from the realm of the gods to the underworld, passing through the middle where the human heart is pinned, representing the necessity of blending sacred with secular.

During my first night in the house, the sky cleared and the

Big Dipper's handle dropped down behind a chaparral-covered peak as if to say that Wyoming, where I had spent many winters, was buried deep inside the mountain and was no longer accessible to me.

At dawn, a white rainbow—a fogbow—arched out of turbulent channel waters and touched the roof. I called Blaine and told him I had not expected to still be alive. He laughed and said, "Of course you're not dead." After, I sat on granite boulders piled in front of the long string of houses as protection against storms. The tides were a form of breathing and the waves were big. In one I saw a single leaf of seaweed suspended in green bubbles, pointing upward, rising as the wave crested, pointing toward light.

My dry eyes filled with tears—not grief exactly, but bewilderment. I was in exile, not in a foreign country, but my own hometown. The wind off the ocean was cold but my tears were heated by subcutaneous fires following the path lightning had taken inside my body, the way ground fires follow tree roots. With binoculars, I glassed the coast north in the direction of Blaine's house, and beyond to Point Conception, the Chumash charnel ground. The stepladder of electricity out of which a lightning stroke is made is the ladder I climbed, fell from, and climbed again.

The bottoms of my feet and the palms of my hands were hot. Everything liquid in me boiled. Now the open window to the outside world from which I had withdrawn sucked in cool fog, viscous as flesh, while my own body was ash riding marine air.

Under a full moon, one western grebe paddled the up and

down swing of incoming swells. The sea was charcoal. Its deep creases surged into waves that knocked the storm wall. During Dickens's time it was thought that coastal people died during ebb tide. " 'People can't die, along the coast,' " said Mr. Peggotty in *David Copperfield*, " 'except when the tide's pretty nigh out. They can't be born, unless it's pretty nigh in—not properly born 'til flood. He's a-going out with the tide. . . . If he lives 'til it turns, he'll hold his own 'til past the flood, and go out with the next tide.' "

I was in luck for a few hours, the tide was coming in. At sunset pink light channeled upward in bright stalks like the aurora. A lone pelican skimmed by as the red globe of sun dropped below the horizon. How could it disappear like that? Would it return? Far out on a western point, three surfers rose and fell as if doing prostrations, as if their devotion could bring the sun back again.

High tide. The stairs to the sand, carved into granite boulders, were wet, and in moonlight the boulders themselves looked like the storms against which they gave protection. I lay in bed with the windows wide open. Where was the grebe? Was she still floating, or had she gone to some safe haven for the night? Perhaps high tide would carry her into my house of glass, glass being nothing but heated sand. I was afraid to sleep because my body felt heavy, like an injured horse who can no longer get to his feet but lifts his head from time to time and begs for mercy. I lay quietly, but each time a wave crashed, the tremor of the house stirred all my thinking.

■ ■ ■

Lightning is thought to be a celestial fire caused by a Greek god's flicking whip, or flashing mirrors manipulated by the Chinese goddess Tien Mu, or stone axes hurled down from the clouds. The image of lightning occurs in all cultures as an emblem of power, chance, providence, and destiny, or as the piercing light that reveals divinity. For Tibetans it is associated with the *Vajra*, both phallus and sword representing the diamond-cutting clarity with which we transpierce self-delusion, preconception, and neurotic fear. Fire represents male energy, whatever is yang—powerful and phallic—while water is yin and female.

Takashi, a farmer-monk from southern Japan who visited me at the beach house, said, "You have always been so strong. Now it is time to learn about being weak. This is necessary for you."

How could I grow strong by becoming weak, I asked. I was being purposefully naïve. What he was asking for was balance. Health cannot be accomplished any other way. I pondered the dampening of the forceful energy, which had always welled up inside me. How does one do such a thing and not ask for death in the process? But that was the point: I didn't have to *do* anything. There was still a lot I had to learn about getting well.

I was weak and the sea was strong, but the swell went flat and pelicans diving for anchovies had to cut up the ocean's surface to make it move again. A single plume of fire shot up out of the sea thirty miles from where I was sitting—not wildfire climbing lodgepole pines, as I had seen in Yellowstone Park, but a controlled stem of natural gas from an offshore oil rig.

Far down the beach a bulldozer pushed sand into a high barrier in front of a row of houses. So much for the acclaimed ocean view. I'd rather be swamped by a rogue wave than wake up to a wall of sand every day. The sky filled with mare's tails—streaming clouds that forecasted a change in weather—and below them, a single yellow tail of smog flew. A squirrel sat on a rock in front of the house, balancing himself by holding a spear of ice plant. He looked out to sea. I looked. What else was there to do? The weather changed but I was the same, or was I?

Intelligence exists everywhere in the body, not just in the brain. An electrochemical pulse beats in every one of our hundred billion nerve cells. It is the "life force" referred to in other cultures. Much like the cumulonimbus clouds, where lightning is born, nerve cells are structured with a difference in electrical potential between the inside, which is negative, and the outside, which is positive, so that in response to stimuli, polarization can take place. Sudden storms of firing neurons travel on long tendrils that sprout from the brain stem and spinal cord and burrow into every organ and muscle in the body. We are the body electric—or, more precisely, the body electrochemical. I no longer think of the brain as being that hard globe atop our shoulders but a body within a body: a long-limbed and flexible apparatus hunting and gathering messages, parlaying them into tiers and nets of neuronal connections almost unthinkably intricate, in which tiny voltage spikes fire into the burning bush of perception, activity, and consciousness.

An electrical impulse travels down an axon—the transmitting arm of a cell—and when it arrives at the axon terminal, certain chemicals, called neurotransmitters, are released from a gated surface—like a gated irrigation pipe—into the synapse, a gap across which those chemicals must travel to reach the other shore. The shore is the dendrite, which is the receiving arm of a cell body.

The synapse is holy. *Apse* comes from *apsis*, whose roots mean, to loop, wheel, arch, orbit, fasten, or copulate, and the apse of a church is a place of honor. The synapse is the gap where nothing and everything happens. Bodies of thoughts swim in the synaptic lake, sliding over receptors, reaching for the ones that live on the other shore. An interval of between 0.5 and 1 millisecond transpires before an impulse makes its way across the gap, as in the bardo where we pause between life and death, treading water in the oblivion of a gray sea.

What is a thought before it registers as memory? Is it a shape, or a shadow, or only unarticulated grayness that can't be held? Is it like unrequited love, or a lover who is spirit only, who has no body?

In the aqueous territory of the synaptic cleft, transmission occurs: the release of calcium activates an enzyme called calpain, which eats into the membrane of the adjacent cell body, changing the shape of dendritic spines and, in the process, creating a physical memory. New neural pathways—new brainscapes—are made each time a memory occurs.

The transmitting axons and receiving dendrites never

touch; it's all in the current, in the gap, in the accidental flow and rhythm of things, in the firing patterns of neurons that decide if a connection will be made.

Patterns of pulsating electrical currents beating rhythmically throughout the nervous system activate cascades of molecular events: the binding of neurotransmitters to receptors causes the release of other enzymes. Some transmitters excite the cells, others inhibit them. When cells die, a group of scavenger cells takes them away. Tightly bound ganglia of nerve cells connected by long fibers are wrapped in myelinated sheaths—fatty, protective tissue. Currents sweep down these telegraphic wires, and in places, impulses jump from one myelin-free area to another, touching bare nodes, called the nodes of Ranvier.

In the midst of this complexity there is simplicity: the electrical alphabet of the nervous system is made up of what's called "stereotyped signals," which are, oddly, quite limited in number and variety. The signals generated in a dog, rat, horse, or squid are the same as the ones that pulse through a human. They are, as Blaine put it, "the universal coins for the exchange of thoughts, feelings, actions, meanings, and ideas." How these simple impulses are encoded and then translated into the structure of experience—what we call consciousness—is not precisely understood. But the complexity is hinted at by the recent findings of neuroscientists: there are perhaps a hundred million interconnected neuronal groups responding to stimuli simultaneously, all the time, in each human brain. And even though neurons that die are not replaced, the dendritic structures are able to regrow, so that in a brain stem injury such as mine, the structures will eventually be repaired.

Thoughts are swimmers that leap, arch, loop, wheel, dive, or dog-paddle in the synaptic gap, the body of water that is like the sea at the beginning of all things, the sea without light. But if living and dying are complementary aspects of the same cycle, then are thinking and not-thinking the same kind of act?

I kept returning to the gap. It is the nervous system's River Styx, where memories, like lives, are ferried. How many crossings do we make in one life? Perhaps the brain is filled with small channels, bodies of water like the one I was living on where Chumash Indians paddled and *aukulash*–a Chumash shaman—sang swordfish songs; perhaps the body is maritime and the act of making memory is natatory—a continuous breast-stroke though a fast current of electrochemical impulses, and the gap, like any sea, is a form constantly undoing itself into a formlessness that rises into shapes through which we can swim. The cleft, the gap, the river into which microscopic channels release calcium, might also be called the bardo of unconditional consciousness, a self-awareness out of which we can't swim.

Fog rolled in like a form of sorrow. To live exiled from a place you have known intimately is to experience sensory deprivation. A wide-awake coma. My marriage was failing and I was not well enough to do anything about it, much less live on the harsh and remote ranch we had jointly loved. Day quickly became night and the fog held. The sea was a memory bank into which everything fell and was lost. I dove in but came out empty-handed. Just before the lightning strike, and for who knows how long after, there was an unoccupied space, a blank I couldn't name or fill. Does memory take place during amnesia, and if not, what does occur?

The fog lifted in the evening and a blue-black band at the horizon marked the end of the sea and the beginning of thought. Where does a beginning begin when nothing has gone on before?

From reading the work of the neurobiologist Gary Lynch, I knew about the architecture of memory, the almost frenetic activity of the brain—hundreds of thousands of connections firing and refiring, tiers and layers of neuronal nets, tissuescapes coming into existence and giving birth to new ones almost instantaneously—but what, I wondered, was the architecture of amnesia, of unconsciousness, of forgetting?

Lynch had found that neurons contain two separate biochemical mechanisms, one for learning and one that goes against learning—a deconstructivist neurotransmitter that breaks up the dendritic structure of memory. But he is not sure what happens during amnesia. Is a memory of a certain event formed, then torn down? Or does some neurotransmitter block the release of calcium and the action of calpain so that no memory takes place? Or are they released but transformed in the synaptic gap by a third enzyme?

To think about thinking is memory in the act of self-creation, causing a new dendritic shape to form inside the brain's circuitry. To ponder the workings of the nervous system—how mental events occur, the infrastructure of memory—is to think about the geography of our psyche. How do we get from a simple, universal electrical signal to a rich conceptual world of imagination, association, and intellect that seems to flow seamlessly as one stream of continual experience? How does such translation and integration occur?

All that's known is this: there is no central processor, no single computer. Nothing that simple. Millions of neurons process information simultaneously and in parallel, not linearly, but the actual chemistry and electrical properties of that integrative process are still being mapped. Even so, it seems odd that during the evolution of brain circuitry and thinking, the ability to understand itself did not get wired in. Such built-in innocence seems like a terrible oversight.

Just before dawn I sat on jagged rocks and waited for the sun to rise. Salt water fell into my coffee cup. Sun tipped the insides of cresting waves as if they were knees bending. Seawater glazed sand into transparencies—window panes that faced a blank. My heart hurt. I lay back against heat-holding granite.

Lazily, I watched a slivered moon move west, its back hunched to winter sun. Maybe I dozed, probably not, but this daydream occurred: I was with friends from the dance company in London. They took my hand and ran until I was floating. All the heaviness, tiredness, and deep aches left my body and inside the fog bank we danced a dance that required no movement yet dissolved mist and dazzled still air.

Wide awake, standing on wet sand in the fog, I found I couldn't really see very much of what was around me. Though my vision hadn't been impaired (some people develop cataracts after a lightning flash), my curiosity had been curtailed. I was too fragile for self-examination. It was enough to get up, get dressed, and assign myself some meager activity for the day that I was rarely able to accomplish. Illness is terribly self-involving—that's how the body heals itself, by appropriating all interests and energies. My vision seemed linked to mental and physical coordination, a way of hanging on wherever I was so that I wouldn't fall. I was still treading water in the bardo.

On Thanksgiving the ocean was royal purple and flapping like a white bird. A band of pollution into which vermillion had leeched made it a painter's sky, a palette whose generosity was to create beauty out of waste. Later the swell went flat. The sea was a void that could not fill itself and continually broke open at my door. When the wind came up, purple water looked tormented, its head all bandaged in kelp, and the channel islands beaten by waves were lost in mist.

Purple changed to blue dappled with pink: the water was a shield, reflecting what was above, below, inside each wave. The waves started blue, ribbed black, ended white. Where kelp beds floated, the sea was bright like ice. Two pelicans skimmed white-caps with their wingtips, and farther out a hidden reef broke waves the way irony breaks open certain truths.

I listened to the healing music of Kevin Volans. A native South African, Kevin studied music in Paris, moved to Ireland, but began returning often to Africa, and his music blends African and European aesthetics. Cover Him with Grass, a collection of pieces resulting from one of these trips, has in it the grandeur and spaciousness of Africa made with sophisticated tonal cadences.

Every morning after sunrise I crouched by my portable CD player and the tiny speakers that went with it and listened. The music worked on me like whiffs of oxygen. I could breathe; I could gallop across wide vistas without having to move; I could look inward with no blame.

chapter 11

Pills can only do so much: they were making my heart beat regularly but they don't give consolation or friendship. On December 16th Sam came to town—Sam being one of the dogs who was with me when I was struck by lightning and who later proved to be a spirit "helper" who guided me back from unconsciousness to waking life. He's my favorite of the young dogs I raised, and to have him with me for the winter was Blaine's prescription for getting well.

Sam is a kelpie—one of the Australian breeds of herding dogs, short-haired, with a wolf-fox face and short tail—and was born on the ranch in Wyoming. He had barely been to town in all his four years, much less Los Angeles. It was rush hour when he arrived in Burbank. When the freight doors opened, Sam's cage was shoved forward onto the sidewalk. I let him out and clipped a leash on. In his whole life he had never seen so many people, certainly not in one place, though he'd seen as many cows.

While waiting for our ride home, we strolled around the airport. Apart from all the cars, now with headlights on, it was children that especially terrified him. Every time he saw a toddler, he rolled over on his back in a posture of bewilderment and surrender. They didn't quite smell like sheep, so what were they, he wondered. He peed on airport shrubbery.

An hour went by. No car. The sky grew dark but the headlights were dazzling. I called. The driver, who was the son of friends, had gone to the wrong airport and it would take him an hour to get back across the city. Weary, Sam and I sat on a bench together, dozing head to head until an airport cop awakened us. "Do you have someplace to go tonight?" he inquired. I said yes. He sauntered by a few more times, keeping an eye on us until my ride appeared.

Up the coast we sped through five lanes of traffic. Sam took in all the sights and sounds he could, then tucked his head under my arm and slept. At McDonald's we bought him a hamburger. "Everything on it?" When I asked Sam, the girl shook her head in disgust and turned away. "He wants french fries too," I yelled.

I raised Sam and nine others in two litters—watched them being born, doctored them through parvo (canine distemper), protected them from attacks by skunks and bobcats, watched them learn to work sheep and cattle. He is the fourth generation of dogs I have known and worked with and is named after his great-grandfather, a black kelpie raised and trained by a sheepherder named Grady. The original Sam was the father of Rusty, Sr., a three-legged dog who was one of the best sheep handlers around (a sheepherder, mistaking Rusty for a coyote, shot the leg off), and Rusty, Sr., sired Rusty, Jr., born at sheepcamp 10,000 feet up in the Big Horn Mountains on a snowy August day and given to me so I would have a working dog.

In Rusty's and my sixteen years together, we endured hard winters and daunting months of solitude. Through it all he developed a quality of peacefulness and dignity, befriending wild rabbits and birds, chickens, horses, and pigs. He was a lover of all creatures except cats, which he hated, and mice, which frightened him. He loved music, favoring the Mormon Tabernacle Choir on Sunday mornings, certain operatic arias, and, most of all, Ravi Shankar. The movie *Gandhi* seemed to entrance him, and he watched it sitting up in a chair all the way through.

Rusty was "tuned in" also to the movements and problems of the animals at the ranch. In the fall and spring, when migrating herds of elk came through and bedded down in our upper and lower meadows, he clawed our arms, whined at the door, and led us to an overlook where we could see the animals, just so we knew they were there.

Once, during calving, his favorite cow was having trouble giving birth. She was half a mile away, the light was going, and it was beginning to snow. Unaware of her problems, we went into the house for dinner but Rusty didn't follow. This was most uncharacteristic. I called for him but he didn't come. Then I saw his silhouette on top of a hill of manure in a sorting corral, where he stood looking toward the lower field. Finally, my husband, Press, went out to see what the problem was. Rusty led him to Spot. By then the calf had been born, half under a barbed-wire fence, and a coyote was gnawing at its frozen back legs. Press chased the coyote off, wrapped the calf's chewed legs in his own socks, and carried him a half-mile to the warming room in the barn, which was piled thick with fresh straw. Because of what we called "Rusty's ranch ESP" that calf was saved.

Too much has been made of evaluating animals' intelligence by their ability to learn human language. Is language the only tool we have for abstract thought? There must be other symbols, ideas, and images that have no names; there must be ways to retrieve them from memory: a smell, a texture, a color, a form, a need flung telepathically across a ranch. Theirs is another alphabet, another string of sounds used to express intuitions and feelings or whatever a dog knows. Just because we can't hear, smell, feel what they do and have failed to decipher their codes doesn't mean they are stupid and we are smart. I'm sure *they* know the opposite to be true.

I sent for a mail-order bride for Rusty. She was fancy and black and, like all thoroughbreds, too high-strung and smart for her own good; but Rusty didn't mind, he loved her unconditionally. Their affair went on for ten days, during which Rusty was a tireless suitor, gazing at her, even as she slept, with adoring eyes. In the middle of ardent and frequent lovemaking, he'd look back at the house, as if to say to me, "You humans think you have fun—look at me!"

Soon enough she had a litter of six: two blacks, two browns, two blonds—an ecumenical assortment—something for everyone. Sam was one of the brown puppies. We gave his brown brother to Grady, who named that pup Sam too, after the original grandfather who started this dynasty. My Sam's spiritual bent was less devout than his father's and his intellect less astute than his glamorous sister's, but that's what was charming about him: from the beginning he was footloose and fancy-free, with a casual, "What, me worry?" look, not because he was disengaged, merely confident that all problems had a resolution.

I watched his mother wean the pups: she turned weaning into a game. Fending the six pups off was the way she taught them how to be aggressive, when to be submissive, how to use their paws and to roll. In all his chasing, rolling, and jumping, Sam stayed on top, not out of toughness but charm and a disdain for repeated failure. Even his anxieties manifested in curious ways: once, while moving cattle off the mountain into the valley for the winter, he stopped in the middle of town and started howling. For five minutes he couldn't be moved as if to say, "I've never been to town before, and I'm not quite sure about it, but I just wanted everyone to know that I'm here!"

Herding and attentiveness came naturally. All those dogs really wanted to do was work and please us. There were no yelled commands, but rather, soft-spoken discussions. "See those cows over there on that hill," I'd whisper, pointing to a few strays. "Why don't you go get them, take them down the creek to a crossing and meet us at the bottom," and off they'd go. All we ever had to teach them was to come back, a kind of self-control that, admittedly, in the first year seemed a hopeless task, but by age two they knew how to move cattle and come back.

Coming from a long line of workers, Sam earned his keep most of the year—except now, during this winter vacation with me in California—so if he wanted a hamburger with french fries, he got it, as all the others before him would have, if there had been a fast-food place nearby.

▪ ▪ ▪

The beach house was dark when we arrived but the moon was bright and we could hear the surf. I called to Sam to follow me out to the deck and down the stairs to the sand. This he did happily until he saw a wave. As it crested, the foam brilliant in moonlight, he ran backwards, terrified of the white water racing toward him, then gave me a hurt look of betrayal. That was enough for one night: he ran for the house and wouldn't come out until morning.

We slept peacefully, his head on the pillow beside mine. Under us, the house shivered with each breaking wave. In the four months since my injury, no one had held me. Now one of my saviors was here at my side: we had both been struck, we had both survived, and I knew that if during the night I fell unconscious, he would bring me back alive.

In mythology, animals are most often the messengers of divine power, and dogs have always had a place in the geography of death. Women are said to be the domesticators of dogs, and in European myths dogs were the companions only of goddesses, guarding the afterworld and helping to receive the dead. In Iran, dogs were allowed to gnaw on corpses before burial. In fifth-century Greece, there was a canine god featured in the Osirian mysteries. He marched in religious processions, standing on his back feet like a human, and when acting as the messenger between heaven and hell his head was sometimes gold, and at other times black. In Asia, the dog was a god called Up-Vat (meaning "opener of the way"), who was said to have started out as a wolf but developed a face like a greyhound, then merged with a jackal, and fed on the dead, swallowing their hearts.

The role of supernatural helpers—guides, ferrymen, or harnessed dogs—stands for the guardian who carries the human spirit forward, whether from death back to life or the other way around. Did my unconscious choice of dogs to aid me come from the intimate living situation I had with them, or was it linked to some collective memory of a time when dogs were associated with funerary customs? In an old Norse myth, the goddess Hel gave birth to wolf-dogs who ate the flesh of the newly dead, then ferried their souls to paradise. I'm not sure if Sam would care to dine on my flesh—he prefers rack of lamb—but he is my guide, my Virgil through these never-ending gaps, these bardos that seem to lie before me.

In the morning after toast and coffee, outside, Sam followed me timidly onto the sand. Nothing smelled right. Where were the horses, dogs, and cattle? He explored, never venturing far, and when he tried drinking seawater, the same look of betrayal came over his face. Can't we even find a camp with decent water? he wondered. He watched the incoming tide carefully, never daring to get his feet wet: these dogs are fastidious, everything has to be in working order at all times, just in case there's a cow to catch. The seabirds went unnoticed.

I knew how he felt. The day I came out of the hospital on my father's arm, the world kept collapsing. There was too much of it, as now for Sam: too many new sights and smells to be taken in. We retreated to the known world of the dog dish and deck chair, like refugees huddled between two strange countries:

the country-of-the-sea-without-cattle and the country-of-too-many-people, and under cloudy skies slept hard and dreamless as the dead.

The next afternoon, at the time of day when lambs and calves gambol and dogs play before the lid of night snaps down, Sam let loose on the beach. His short tail tucked under, back feet wheeling up almost to his ears, he ran in great loops and wide figure-eights. Finally, far down the coast, he veered off to chase a sanderling, then, turning and running in my direction, he leapt full tilt into my arms. Home.

chapter 12

The French call the relationship between doctor and patient "*un couple de malade*," meaning paired-up, yoked together by illness, like a marriage. It is so. There is chemistry between the healer and the one being healed, and those who minister to the heart evoke a profound and tender connection. To be "yoked" like a pair of oxen is an image I liked, because it implies the dual effort it takes to get well or else die properly.

If your heart stops, you have four minutes to get help or resume spontaneous rhythm before the brain stops functioning properly. Who is going to get the gift of the next heartbeat, wonders anyone whose heart has stopped. As Blaine's patient, I was hooked to him and he to me by that exhilarating urgency.

We argued about whether or not he saved my life. I said he did. He said I just needed someone who took the time to see what was going wrong. But the healing has been chemistry. He sat on my hospital bed when he talked, shooting a million questions at me, not from the pathological premise of illness, but from the sane ground of surviving well. He asked about work and love, as well as about dizziness and chest pain; he charted a course for how I could live given my disabilities, and the course was vaster than what I could have achieved just then—it was something to aspire to. "Do anything you want," he said, "as long as there's no chest pain. You can always lie down if you start to pass out. That might be a little awkward sometimes, but who cares?" Death didn't daunt him, nor did the inconvenience of injury.

Two thousand years ago Hippocrates said: "A patient who is mortally sick might yet recover out of belief in the goodness of the physician." About the patient-healer relationship Blaine said: "If you care about your patient, that caring will dictate your behavior; you'll listen, ask questions, listen again, and ask again. The patient will tell you what you need to know."

The thickness or thinness of a doctor's armor determines the distance he can see into a patient, the intimate ability to touch the vital force surging somewhere through the body, undetectable by MRI, CAT scan, or angiogram. Blaine put his stethoscope down on my heart and listened to forty-six years of thumping—joyful detonations and solitary longings—just like everyone's. When he lifted his head and opened his eyes I thought he must already know a great deal about me—the flawed toughness, nicked by fears and loneliness, and the excesses of passion—and that he could see if Death was still in the corner of the room, or at the end of my bed, but he said nothing.

Above and beyond the drama of cardiac arrest, or the threat of it, is the metaphorical territory of the heart: if love desists, if passion arrests, if compassion stops circulating through the arteries of society, then civilization, such as it is, will stop.

When I looked up at Blaine from my hospital bed after my heart rate and pulse had dropped out of existence, he appeared to be some kind of all-American bodhisattva, someone who travels the middle way that is the human realm of the heart, not because he holds some tantric mastery but because he is so thoroughly human: bumbling and precise, brilliant, corny and sane, optimistic and ordinary. Perhaps the tenderness I felt toward him—which he allowed to happen because he knows that a doctor who is arrogant and thick-skinned will heal his patients only as well as the technology works—was also a tenderness I was able to feel toward myself.

After a slump, I felt almost maniacally exuberant. One asks, "Is this really me, alive?" as often as, when on the brink of death, "Is it really me dying?" In the examining room, Blaine took my blood pressure. It still dropped twenty points when I changed from a supine to a standing position. I was taken to a room where a stress-echocardiogram would be performed—a test in which the heart muscle is looked at by means of a sonogram immediately after strenuous exercise.

When Blaine put me on the treadmill my feet almost flew out from under me and he grabbed me around the waist. The nurses howled with laughter: "Dr. Braniff, you've got the treadmill going about eighty miles an hour!" one of them exclaimed. His face reddened with embarrassment. "So much for confidence in your doctor," he said. "Do you do this to your ninety-year-olds?" I asked. After ten minutes on the treadmill at a greatly reduced speed, I lay down and the echocardiogram was taken, while I was still sweating and puffing. He looked at the image of my heart working hard. "You have a wonderful heart!" he said. "Great contractibility. A twenty-year-old's ventricle."

Is that as good as having great legs?

Doctors can mete out death sentences as surely as they give medicine. If they say "you have six months to live," you believe them, and sure enough, you die. One of Blaine's gifts is a belief in the resilience of the human. His ardor is contagious: it's made of his cockeyed optimism and our own vital force. However weakened by illness, a thread of vitality pulls through: his vitality becomes ours and we revel in its presence, no matter how long it lasts: an hour or fifty years.

We are elaborate biochemical, electrical, emotional organisms with message systems so intricate no computer could begin to track what happens to the body when even a single thought registers there. Feelings change the chemistry of the body as surely as physical traumas do. We blush, we faint at the sight of blood. Cells are constantly sending messages and reacting to messages from neurotransmitters. The mind-body split is a meaningless, laughable idea. Neurons are strung along electrical paths like Christmas tree lights, dancing and blinking, tiny intelligent beings that illuminate the dark continent of flesh. Neuropeptides and hormones are released, chi flows, surfacing in points of electrical resistance, where needles are inserted and vibrated to unlock obstructions. Each nuance of emotion makes its mark and every physical shift is met with a homeostatic adjustment. We sweat, change our heart rates, urinate, vomit; we mend bones, keep our kidneys working, alter heart rate and blood pressure, pursue enemy viruses, change our blood-sugar levels, and respiratory rates.

In 1859, a French physician, Claude Bernard, said: "*La fixité du milieu intérieur est la condition de la vie.*" Perhaps homeostasis is the body's greatest accomplishment: its mastery of maintaining a constant internal environment despite constant and drastic external changes. From it we learn what the great spiritual teachers have already tried to teach us: that stasis is achieved through dynamism; that constant change is a form of equilibrium; that to be ordinary is an outrageous extreme; that limitation can be freedom.

chapter 13

On December 21, no one mentioned the winter solstice. By "no one" I mean the nurses at the doctor's office or the young man who came to wash the windows at the house, since they were the only people I talked to that day. While the earth was working on its solstice balancing act, the flow of electricity inside my body was not. Instead of taking my usual walk, I tried to stand perfectly still without losing consciousness. An extreme high tide broke over the seawall, spreading white foam at my feet. At the moment of the solstice, when the earth shifted into its axial equipoise, I fainted.

When I could see again I saw dolphins arching by. Dolphins bring luck. The appearance of a fair-weather mackerel sky seemed like a conciliatory gesture in mid-winter: a delicate balance between wet and cold was held, but a balance maintained too long dies. California had endured an eight-year drought. But later that day, storm clouds blackened the West. Rain let down north of San Miguel, right where the sun sets at this time of year, but on the beach where I stood with an aching heart and clammy hands, sun prevailed.

Christmas came and went. Of it I remember little except that when my parents and sister came bringing dinner on Christmas Eve, I could not even manage to set the table right, and in the morning, when I tried to bake a chocolate cake for my father, in thanks for his lifesaving flight to Wyoming, I failed.

On New Year's Eve the chest pain and light-headedness was so bad I went to bed at eight. Sam's body was pressed against mine. Did I really need anything else, anyone else in the world? At midnight I woke, stirring to the faint noise of firecrackers, looked at the clock, rolled over and kissed Sam. I wondered less about if I was going to die than if I was already dead, but my body told me otherwise: my whole left side felt as if something had been detonated there. I wanted to be held, to be pieced back together and fastened to the realm of the living by another human being, but there was no one, and there would be no one in the morning.

Sam sighed deeply and lay on his back, feet in the air, teeth showing. Waves rocked the house, which rocked me. At dawn a slim moon floated above cerulean water, and bright Venus was the pinprick of promised love that had failed to appear. In my version of things, the new year would begin not with drunken abandonment but with burning: great bonfires up and down the beach, into which people would toss the detritus of the old year—relics of habitual thought and spiritual materialism. Then there would be food cooking over flame: meats and cabbages, potatoes and corn, as we nourished ourselves with the new emptiness, and, finally drinking and dancing as high tide came in, until every bonfire had been drenched by a wave, we would dance in water, underwater, nearly drowning—until the dogs came for us and swam us home.

New Year's morning I stood on rocks and burned a photograph of a man who had jilted me but to whom I clung as a friend out of hope that his affections would rekindle and fear that I would end up alone. It was not him I was burning but my own habit of attachment.

The tides know everything about habit, but also everything about cleansing and healing. Ash from the photograph fell into the foamy edge of a broken wave. Sam jumped sideways to keep from getting his feet wet. I was sure he would have laughed at my clumsy ritual if he could have, so I laughed for him. Then we walked to the estuary, where a Chumash village once stood, its hunchbacked huts made from whale ribs covered with a thatch of reeds. Having risen from my ashes several times, I found there was still more burning to do, more fresh starts to make, but before long, I felt too tired to contemplate such enterprise. We were well into New Year's Day, and the starting line had already been crossed.

I was blind, or at least I felt that way. The calendar was blank, no numbers to differentiate days, and I still couldn't read, couldn't use fictional structures to scaffold my interest in life. My friends, Noel and Judy, came and took Sam and me out—I wasn't allowed to drive. Even though I had lived through Wyoming winters, the California nights felt cold, and they built a fire. We broke bread, ate soup, drank red wine. At the movie theater I carried Sam under my arm like a lamb. They said, "You can't bring him in here," and I said, "He's blind and I'm his seeing-eye human," a joke they liked, so they let us in. We saw a Mongolian film whose wide open landscape reminded me of Wyoming. It was almost too much to bear.

On the ranch, Blue, my seventeen-year-old sheepherding horse, had often come halfway into the kitchen looking for dry dogfood, which he liked. His head was so big, it threw everything else in the room out of proportion. The dogs went crazy with delight as he stuck his Roman nose into the fifty-pound bag of mini-chunks and basked in their loving gazes as he ate.

Coming off the mountain, I'd sometimes let Blue take any route he wanted. It was a way to see the map of his mind. He liked to smell pine trees and eat purple thistle flowers with puckered lips. On cold days he'd choose the "steep trail," which was almost vertical, though it was the fast way home, but on warm summer days he'd use the trail that took him through old hunting camps, because he liked to smell the places where horses he didn't know had been and he could drink from the creek that flowed through.

Later Sam and I stood on the seawall in front of the house. I didn't want to think about the ranch. The continual sound of crashing waves was a vise that held my thoughts in the present. The beach was empty and the houses were dark. An acquaintance who spends time in Labrador said that for help during a hunt, the Inuit people placed bones in a fire and when the bones cracked, they formed a map of trails on which to find caribou, and that as a result, there had been a lot of hunger there. I wasn't hungry—in fact, one effect of the lightning was to dampen my appetite almost completely, but, living in exile from the only place I knew intimately, I needed a map, I needed the oracle of bones.

In the morning I followed Sam's map of smells to rock piles, tide pools. He had been timid at first, then grew bold: the beach was his domain. I tried to imagine how his olfactory sense opened every inch of ground for him—successive explosions at each step, widening into a map of the world, but what a map: guano, salt, fish scale, seal fur, tar, and the mineral smell of sand. The beach was near the town of Carpinteria, named by Spaniards for the Chumash Indians they saw there—the carpenters—who were building their seaworthy canoes called "tomols," sealing them with the asphaltum that still oozes from a nearby cliff and piles up into a hill of black bubble gum on which sea lions bark and lounge as if waiting for the tomols to return.

Though Sam is no bloodhound—he'd rather look at a band of sheep and lick his chops than smell the guano of any seagull—when the water went out, his nose took me to the edge of the shore, to exposed rock. Tide pools are another kind of gap—an edge between batholith and lithosphere, ocean and earth. They are ecotones, in-between places like those clefts in the brain and the rug-pulled-out limbos in our lives where, ironically, much richness occurs.

Down boulders, across sand, between clumps of kelp,

Sam's paw prints in wet sand were dark asterisks on the map marking the trail. That day the beach was a bed of black rock hung with the slanted roofs of barnacles, mussels, and the limpets' pink volcano shells, and pocked with smooth basins catching the green splash of waves. Each pool of water held an image of the sun; each sun was a lake of daylight, and when the shine of the rock started to fade as it dried, another wave splashed it bright.

A tide pool is a kind of meadow: rocks bared by low tides are strewn with red algae and green mermaid's hair—seagrass, surfgrass and eelgrass—which is brushed back and forth in undulating waves. In the splash zone, village life seems to prevail: both shelled and soft creatures hide under rocks, between rocks, or fasten themselves to rocks in colonies, rigid in sweeping waters and in the continual flux of tides. Barnacles, anemones, jellyfish, starfish, sponges, hydroids, worms, chitons, mussels, clams, snails, octopuses, shrimps, crabs, lobsters, sea spiders, urchins, sea squirts and salps, algae, kelp, lichens and grasses—all crowd together in urban densities. Aggregate anemones often live in concentrations of three thousand individuals per square meter of rock; they are marine apartment dwellers.

A tide pool is perhaps the one place where creatures can prosper by becoming completely sedentary and permanently attached. Barnacles spend their youths floating free before coming home to a rock and, once glued, they never leave again, never have to go in search of food because room service is provided by the waves bringing it to them. They merely extend legs to capture floating plankton; these legs are also the apparatus through which they breathe. Mussels are "gill-netters" They catch prey, pump water, and breathe through fleshy tubes that extend from the rear of their hinged shell.

Because sessile creatures—ones that are permanently attached—are unable to "go with the flow," they thrive on extremes: they're either underwater or desiccating in hot sun. To avoid drying up during low tide, periwinkles and limpets keep so tightly closed that their respiration stops. They literally hold their breath until covered by water again.

A regimen of daily walks took me into the middle of January. With each high tide more sand was removed, more rocks exposed, more sea creatures revealed. In dense fog a single red starfish, bright on wet sand, was the only thing visible and seemed to stand for the whole unseen galaxy. Up close they're less than romantic. If barnacles and mussels represent tenacity—not only are they unmoving, but also very long-lived—then starfish are known for their voraciousness. Known as the "walking stomachs of the deep," they eat continually, devouring everything in their path: oysters, clams, barnacles, and they sift through sand and mud for any bits of bottom garbage and carrion.

A starfish eats by pushing his stomach out through his mouth, located at the center of his body. That's how greedy it is: it eats everything whole, shell and all. And just so nothing is missed, the branching tubes in the arms, where its eyes and respiratory system are located, are put to work picking up any dropped morsels. The efficiency of the starfish's stomach is twofold: waste products are ejected from the same opening through which the meal was ingested.

Brittle stars, blood stars, ocher stars, leather stars, variable and shallow-water sand stars—the ones seen around here—as well as 3,600 other species, all start out as one of thousands of eggs shed from the underarms of a mother/father (they are hermaphroditic). Once released, they glue themselves to kelp. From "stardust" they turn into odd-shaped punctuation marks until the rays finally emerge; they grow from stardust to gluttony in a matter of weeks.

Sea stars, as biologists like to call them, since they're not fish, have an extremely complex nervous system and are famous for regenerating lost arms. They can cast off a wounded ray and regrow a new one, or else grow a whole new body if part of the central disk is included in the cut-off arm. With such powers, starfish numbers could get out of control, but this problem is solved by sex changes—no operation needed. Males can turn into females, and vice versa, an art that regulates population dynamics.

As I poked through tide pools, Sam began stalking shorebirds. He never hurt one, rarely even chased them, just sneaked up on them, wishing they were sheep or cows. Sandpipers, sanderlings, phalaropes, plovers, godwits, curlews, and pelicans flew in front of him as he ran, circling around over the water and landing behind him. On one of those days I stumbled on something shaped like a tiny leaf, though it felt alive, like an animal in my hand. Animal or vegetable? Less than half an inch high, soft like a cut-off earlobe, with a purple stem, tiny volcano-shaped pores, and a calcium carbonate "skeleton" made of microscopic spicules, it later proved to be a white fan sponge, severed from its other friends: the urn, macaroni, and crumb-of-bread sponges, the purple, yellow, orange, and free-living sponges.

Sam's paw marks and my footprints crisscrossed that of a snail, which secretes a mucous trail to trap plankton and small creatures, then reels in the trail to feed on the catch. We came upon spotted nudibranchs that looked like bits of body parts, unshelled, skinned, globs of flesh exposed to sea and air. They were easy to spot near the sponges, because that's what they dine on.

Nearby were brown sea hares, which look nothing like a rabbit but are in the mollusk family—a type of snail. They lay their eggs in long strings that get wound up into spongy yellow balls the size of grapefruit and contain as many as a million eggs apiece. When the larvae hatch, in ten days or so, the young swim free, but most are quickly eaten. This is rather fortunate because if they weren't, the population of brown hares would exceed the combined populations of animal life on the entire earth in only two or three years.

Fertility, gluttony, tenacity, and a who-cares-who-does-what-to-whom sex life—that's what typifies creatures of the intertidal zone. No sinner has had a life as rigid and voluptuous as the tube snail, which lives clustered and unmoving, its shells growing entwined, or the sea cucumber, whose water and oxygen is pumped in and out of its anus, or the vaginal-looking aggregate anemone, which stings its prey with a paralyzing toxin and reproduces by splitting in half lengthwise, generating two individuals of the same sex.

chapter 14

A black-crowned night heron stood on an apron of wet sand, looking across the channel. The feather plume at the back of his head lifted in a faint breeze. Out there the channel churned its cyclonic eddies counterclockwise. Schools of anchovies, halibut, and sea bass came and went: silver flashes, small storms that well up from the inside of the sea but are short-lived, like lightning.

A shifting wind sailed up the beach from the southwest, which means rain. As he walked, sand blasted Sam's face and ears. Spray from the tops of waves blew backwards into the sea and came up again as whitecaps, and the green shoulders of swells pushed hard toward shore but never seemed to arrive. From the beach I could see the plate-glass windows of my house bulging into wide-angle eyes, and wondered what they could see.

A hundred and fifty thousand years ago, southern California turned frosty during two ice ages, beginning a long warming trend that continues today. The sea here was 400 feet lower than it is now: the channel was a mere lagoon, with little surf, and four of the five channel islands were fused together into one large ocean-bound mountain range. The mountains behind

Santa Barbara had a covering of pine forests and only in later, warmer years did they change to oak savannah. Despite the mild climate now, this is one of the roughest channels in the Pacific. Two currents, the warmer Japanese current—Kinoshiro—and the cold California current, meet and sheer off at Point Conception and on the northwestern tip of San Miguel Island, and prevailing winds bounding down the coast funnel through the channel at high velocities.

Sam and I climbed a rocky knob and tried to see out beyond the protective arm of the islands. "If you look hard enough, you can see the Antarctic, Hawaii, or Japan from here," I told him. But I wanted more: I wanted to see the topography of the ocean floor. Ocean covers 78 percent of the planet and averages a depth of 12,000 feet. Its topography, a frontier now "seen" by side-scan sonar, which produces high-resolution images of the ocean floor, is as grand as the terrain up top, but there's more of it, since the volume of habitat in the sea is much greater than that on land.

I thought about all the ways the ocean covers things. Rogue waves can appear suddenly, as if the sea floor had shifted its buttocks, causing an ephemeral ripple that swallows fishing boats, islands and people. Instead of a tsunami I saw only pelicans bobbing on swells that broke into tame, three-foot waves.

In the evening we went for another walk. Sinking into sand, Sam's tracks were bright meteors, appearing suddenly, then fading to black. I stood in water up to my knees, grabbing phosphorescence and invisible plankton, squeezing light out of the ocean's dark brew. Light is chemical, electrical, mineral, just the way memory is, and I wondered if light had invented the ocean and my hand dragging through it, or if memory had invented light as a form of time thinking about itself.

Down on hands and knees, eye-to-eye with Sam, I tried for an all-inclusive panoramic view. Between tide pool and sky there were rhythms and plasmas holding me up. Lightning, cosmic dust, human blood, ash, plankton—these are all referred to as "plasmas" by scientists, as "broths of life" that rocked and rolled to various kinds of music along with the fluctuations of the dinoflagellates periodic light. Did they have the same beat as the electrical firing pulse of neurons in the brain, or did spring tides (those at the dark of the moon, or when it is full) and neap tides (when the moon is in between) match the shifting and quiescence of tectonic plates, or does everything modulate only to chaos?

The next day the heron was back. In all those rhythms he stood motionless, the solitary observer of a navy sky graying with rain clouds. Between two rocks a starfish righted itself by lifting up on a single ray and flipping over. Earlier, a painting by Picasso had caught my eye: it was a blue acrobat rolling like a wave. I thought of that figure as the blue acrobat of time who had given me a reprieve from death, had lifted itself up on one arm like the starfish and let me slide through.

The night heron's white plume moved stiffly in wind like a compass needle: to the west, northwest, then south and east.

The sea looked like time, and time was water and tides, the heart's ardent tick and the sea star's flip. Spreading his gray wings, the heron rose slowly: the storm he had been waiting for had come.

chapter 15

A sound of tearing ripped across the tops of waves like a torn seam. Rain was the thread from which fluted waves had come unstitched. Wind poured down. Lightning cut sprays of rain that fell and lifted and seemed to emanate from the sea, not the sky. Palm tree fronds swept together in green flags and an oriole's swinging nest made of palm threads swayed with it, carrying unhatched eggs. A wind from the east blew seawater against its own current. Shorebirds hunched down on driftwood and pelicans flew in threes over the roofs of houses instead of shaving the tops of swells with their wings. What had been the waves' transparencies, through which I could see feather boa kelp and ducks' feet paddling gracelessly, were now walls of water muddied with silt. Every few hundred yards another creek carved an urgent route across sand—water thirsting for water—and at the confluence, great fans of suspended earth spread out into the sea.

Immersion in water stands for annihilation and a return to formlessness, followed by rebirth and regeneration. In the geography of death, it's not the ferryman who counts but water itself:

the ocean as medium between fire and air, life and death. After lightning struck I felt like a twig in the shape of a cross floating inside the sea.

When the storm came the whole world began moving. Brown ribbons of kelp came untied. Swells that rose tall as buildings sank into black troughs and the troughs redoubled themselves so that the swells kept falling. Rain shot down from all directions at once, then it was a silver sheet sliding into the sea like glass, pane after pane shattering. I left the doors to the house wide open to welcome the storm, secretly hoping the waves would come in. I'd dreamed a tsunami had broken through windows, that I'd stood in a shower of foam that rained down people, dogs, and horses. After coming so close to death, these were my quiet celebrations of life.

A blue hole in the roof of the sky appeared. The ruins of the storm washed up: lemons, plastic bottles, oak limbs, avocados, bumpers from the side of a boat, a fiberglass hull. In four days, five vertical feet of sand had been taken away, and exposed rocks rattled in surging tides. From the beach the row of houses looked taller, as though the moisture had made them grow. Out beyond the waves that came in series of four and five, not one or two, a pelican dove down into kelp beds. Each time, the sea closed over the bird but let go fast: he rose and flew. Floating logs pretended they were seals, and seals poked their heads through the middles of waves. Surf scooters—homely diving ducks with black feathers and thick red beaks—floated en masse, resting before continuing on to Alaska. A wreath of black clouds that had broken away from the storm pulled apart over the islands, but no sunlight poured through.

During the storm the boat of a local fisherman went down. He had been fishing for halibut up the coast when a wave swamped the foredeck and bridge. The wheelhouse filled fast with water. Barefoot, he had to kick out the windows with his feet to escape, and as he did so, saw his dog tumbling, then she was gone. There was shark danger, intensified because of his bleeding feet but his coworker—called a "tender"—had plucked two survival suits from the sinking vessel, and when they finally found each other in churning water, they managed to get the suits on. Though survival suits won't prevent shark bites, they'll keep you warm, even if you're already wet. When they didn't show up that evening the other fishermen called the Coast Guard and the two men were rescued after being in the water for five hours.

The next week, sand-bearing waves brought the beach back, erasing all the scars of the storm. The coastline kept reforming itself—revising drafts of how it should be shaped, how many rocks, how deep the sand. Only the unmoving tide pool creatures stay the same.

One morning the sky was a red wall—meaning fair weather—and at dusk it was an orange flame. But water puts out fire and no harm came to me. Except that unexpectedly I had to move, because the owner of the house I was inhabiting wanted to return. That house had been my refuge and the sea my restoration. I had a week to find a new place and I was too frail to pack and lift boxes, to move anywhere.

chapter 16

The Arctic people of Labrador say that a person is born empty: dreams fill him, and a person who doesn't dream is no better than a black fly. That's what I was, because I'd stopped dreaming almost completely since being hit by lightning. It's now known that REM sleep is associated with a surge of sympathetic nervous system activity—of which I had very little, and so for six months my nights had been empty.

Then I dreamed that I died. Water surrounded me and nothing looked familiar. On waking, I found that I hadn't died, only moved. From an austere minimalist house on a lonely beach to a small cottage in a cove where the residents were gregarious and walked their dogs every day: I had been reborn into the realm of humans.

Instead of tide pools, Sam sniffed trails to other dogs. We met Skippy and Minke—two Skipperkees belonging to a tall, long-legged, world-class diver, writer, classical pianist, and adventurer named Hillary. We met Thatcher, a huge German shepherd belonging to Kate, a glamorous and vampish Australian who might have starred in Cecil B. DeMille's classics, and down at Fernald Point, another dog, named Dennis, who liked to chase boats from the beach, a habit Jim, his owner, bemoaned: "Why can't he just chase balls like other dogs?"

In the other direction there was a standard poodle, two black dachshunds, and a variety of surfers' mutts, but Sam liked Dennis best, perhaps because their coloring and markings were almost exactly the same, though Dennis was larger.

On days when I felt too dizzy to walk, I'd send Sam down the beach to visit his friends. "Go find Dennis and bring him up here for a snack," I'd tell him, and off he'd go, a quarter of a mile down the sand, up through Jim's elegant garden. Sometime later the two dogs would appear, looking very pleased with themselves. Or else Jim would call and say, "Sam's in my kitchen, do you mind? I'll send him home in a while."

Because Kate was homesick for Australia and the rituals that go with home, we decided to have a weekly Wednesday night "barbecue," to which all the dogs and their owners were invited. She grilled "snags" (sausages) for the dogs and for us, and I'd arrive at her house down the beach carrying a wooden salad bowl—on my head or hoisted up on my shoulder if it was a high tide. I liked the fact that we'd all met because of our dogs; dogs don't care who is rich or poor, accomplished or struggling.

The nights were balmy. "Wet Wednesdays" were race nights at the harbor and the horizon filled with bright spinnakers. The dogs played until darkness came, and in the sweeping peace of the tide going out, they lay down all in a row and slept while we drank to the blessings of friendship, canine and human alike.

■ ■ ■

It was spring, and while my new house was cramped and humble, it was on the sand and the ocean still came to the front door. At dawn I'd roll out of bed, not even bothering to change clothes, and walk. Squalls came and went. Storm surges carried huge swells into the cove, and as rain inebriated the coast, the thick stub of a rainbow pushed out of the sea like a green thumb on the horizon. After, the dark blue sheet of water turned metallic, and I wondered: What in nature is not a mirror, does not give back a true image of mind?

It was March, seven months since the accident, but not without setbacks: when Blaine took my blood pressure, there was barely enough cardiac profusion, meaning my blood pressure was absurdly low. What had seemed like steady progress toward health was a fiction. There was only shuffling forward and leaps back. Just as capriciously, two kinds of winds started blowing: the sundowner, fierce winds that came from the north, funneling over the coastal range like a snake, then blowing out to sea; and the Santa Anas, hot desert winds moving in from the Mojave Desert.

Between wind storms El Niño hoisted the jet stream on its humped back, bringing tropical moisture on narrow clouds stretching all the way across the ocean from Hawaii. The rains were monsoon-like, dropping two or three inches an hour, bringing an end to a severe seven-year drought that had downed 300-year-old trees. The bright crenellations of sea and the cliff faces of distant islands were like false maps of places I might go to if I could have walked on water or slipped out of my skin. But at least I wanted to go, even if I wasn't able.

Late in the day all that was left was a sliver of light on palm tree trunks and the stems of fronds blunt-cut like hair. For so long I hadn't been able to see; that is, I didn't have the energy to absorb what was around me. Then things began to interest me again. After all, it was spring. I told Blaine I thought my cortex had turned green because it started giving me dreams: I was swimming upstream, leaping up waterfalls to get to a spawning ground. But someone—a huge human figure—was obstructing me. With legs spread, he stood above me, throwing me out of the river as I approached, but I leapt back in again.

Sometimes I walked the hills overlooking the ocean to get away from the drone of surf. I walked slowly and relished the return of migrating and mating songbirds. I wanted to walk when I could and it occurred to me that stillness doesn't mean not moving—seated meditation is only a reminder of a quality of mind in which one is wakeful, lively, spirited, humorous, not acting out of desperation.

At my new house a green symphony played: clumps of volunteer bamboo clacked so loudly it might have caused whole orchards of lemons to ripen, and woodpeckers drilled into palm trees. Dreaming had loosened the terrible claustrophobia I felt as a result of amnesia and six months of blank nights: it knocked peepholes into the mind's closed rooms.

Blood orange and bitter lemon trees blossomed, rafts of migrating ducks flew north in ragged, aerodynamic formations half a mile out from shore, their many wings appearing as one, and swallows arrived to build mud nests on sea cliffs with an opening at the front so they could take in the view. Walking was my slow-motion flight from and back into civilization, my meditation in action. When afternoon winds came, clouds that rose above the islands were shaped like the islands—mist imitating litho-sphere—and hillsides of eucalyptus bent their fragrance down to wild, whitecapped seas.

chapter 17

In April a dead seal washed up in front of my house, and a duck, and a Christmas tree. Where storms had stripped the beach of sand, exposing rough cobblestones, brown waves brought sand back, erasing all evidence of death, and in doing so, provided a fresh canvas on which to etch our comings and goings. Then other storms came, disinterring what had been hidden by sand.

Up and down this small beach, madness had not been a stranger: a house built on the edge of a Chumash burial ground burned; a woman's young lover stabbed her husband to death; a doctor put on his tweed coat to go to work but instead walked into the sea until it closed over his head.

New sand makes the beach a graveyard, sweeping over whatever was there a moment before, and the tides erase even that. And no matter how far a foot presses into the flesh of the earth, it is pulled from it again by water.

A beach is where the rock of the planet is ground down into minutae, where the general is splintered back into particulars only to become one thing again—a collective body of sands. Our human and animal bodies are mostly water, as is the planet, and water eventually takes everything and is everything—the true corpus whose aqueous flesh remains after bones have slipped from the envelope of skin and feathers have separated from wings.

This is the beach where I first touched a boy's penis during a makeout party, listening to Elvis and the Everly Brothers, quickly trying to figure out in the dark and under layers of clothes how it uncoiled and where it went when it stiffened. Now the house where we had those parties is closed up, and near the one that burned down a jacaranda tree blooms in cascading fireworks of red flowers. Beyond, dark creases in the sea opened and I saw the back of a migrating gray whale.

These intimations of spring were accompanied by a deterioration in my health. I had trouble staying conscious even when lying down. Blaine was in the Galápagos for two weeks and I decided to tough it out until he came home. One morning a painful knot in the center of my chest tightened. A dull ache traveled down my arm to the elbow, my heart resting in the elbow's bent L, trying to spell out something, maybe the word *love*, or maybe *lost*.

Chest pain has its own particular geography; it is territorial, rising in the center of the upper body; the heart is a taproot laboring to bring minerals to the surface. Sometimes pain wraps all the way around the back in long vinelike lianas pulling tight, and underneath, the heart is an island floating—maybe Greenland, or an aquifer with a tentative land cover like Labrador's, which makes me wonder why the equatorial waters at our centers, the punched hole of the naval or the muscular knot of the solar plexus, aren't thought to be the seat of love.

▪ ▪ ▪

A voluptuous season of mountain wildflowers came and went, and I missed it. While trying to take Sam down to the beach one night, I passed out on the rock steps in front of my cottage and sprained my ankle. After that, I gave up even the least adventure. For ten days I was the sea's prisoner, a sea that looked swollen and hard, a single oscillating block. Each evening the sun slipped into its hiding place, which the Chumash say is the hole in the top of the sand dollar—and rested in its round, gold room.

The moon was a ghost that rode the ocean at night. Waves were the deformed prodigies of a marriage between moon and shore: they were the pages of an unwritten book. Wind scuttled white foam until it tore—an alphabet coming apart. People asked: "Are you still able to write?"

The aboriginal people of Taiwan thought there were originally two suns, but one was shot down with an arrow and became the moon. It might have been more fitting if the sand dollar had been designated as the resting place of the moon, since they reproduce once a month in summer on nights of the full moon. Such amorous and romantic creatures! Thousands of eggs are released into the sea where they float until they are found by sperm of their own kind: a rendezvous which is accomplished by special "recognition molecules,"—knobs of protein that jut out from the egg's membrane and link only with sand dollar sperm but no other.

I groped in darkness and found I was bound to no one, by no one, and possessed no means of reproducing myself. It was hard to hold my head up, and the air around me seemed dark.

The sea was a cauldron of passion and intimacy, which I hated that week. How maddening water's fluency was as I lay transfixed by dead brain cells. Night after night I wondered what would become of me, and hoped that whatever it was, it would come fast. I sat in my dull cocoon and watched waves break into formlessness, then form back into waves again.

In one dream the whole sea turned into human flesh and was divided up among shells. Shells were ears into which salt water sluiced, and out spilled the driving sound of surf onto my pillow, a sound so loud it woke me: I had been crying. A lump in my throat rose, a small planet shaped like my ranch in Wyoming: pastures, meadows, house, barn, sheds, lake, all rimmed by high mountains. I spit the lump out and it floated on a loose ocean, drifting until the horizon took it away. When Blaine returned from the Galápagos, my spirits rose, as did my blood pressure.

At our next Wednesday night "barbecue" Kate, my friend who lives in semi-exile from Australia, broke down in tears. We had been talking about northwestern Australia, where alligators eat illegal Chinese immigrants, and about the high, pristine Snowy River country, and she said, "I'm not all gold jewelry and makeup. I miss my land. I want to go into the bush for a month and walk around." I cried with her.

The next day we met while walking our dogs. Suddenly Sam started bleeding from his rectum and doubled up on the sand. I ran to him. Kate was already heading for her house and yelled: "I'll call my vet. He's the best in town. He'll be waiting for you." She explained where his office was as she took off. I carried Sam down the beach as fast as I was able, through my house, into my pickup, and speeded—lights flashing—into town. As we were driving I thought that if he died I would not be able to go on without him, he was my last hope, the thin thread that fastened me to all that I had known and loved and lost in Wyoming.

The vet, tall and quick-minded, examined Sam and immediately started an IV of glucose. He fired questions: What had he eaten, where had he been, what were his habits, had he been sick before? An X-ray showed no intestinal or bowel obstructions, and a blood test eliminated poison. "It's got to be a wild bacterial infection."

Sam lay motionless. His flank was caved in and his breathing was labored. I held him while the nurse shaved another patch on his leg for a needle, and another IV—of antibiotics—was begun.

Later, after I had done all I could to assist and Sam was sleeping, I went home to rest. How quickly my own health problems had been reduced to nothing. My chest hurt but I didn't care. All I knew was that I could make it to the vet's office from my house in exactly six and a half minutes, and that Ron, the vet, had been instructed to call me if anything happened. All my thinking was for Sam, about Sam. During the evening, I visited him several times. Ron was there because his own dog had also been hurt—hit by a car—and so he attended to our two dogs through the night.

In the morning Sam was better but they kept him another day. It pained me deeply to think about his confusion: What was happening to him, why was he in a strange place, why were people hurting him? When he came home finally, he was very thin. I fixed his prescribed bland diet but he hardly ate. Instead, he curled up on the bed and slept with his head pushed against my stomach. Every few hours during that first night I turned on the light to make sure he was still breathing and that the bleeding had stopped, and held him in my arms until dawn.

Then it was Sam who couldn't walk far down the beach and I adjusted my speed to him. Some days, neither of us could go very far. But even if we did nothing but sit on the rocks, we saw things: pairs of western grebes, with their slightly flattened heads, floated in water on roller coasters of rising waves, and Sam thought they looked like rattlesnakes that could swim.

He was too weak to chase birds, so I taught him the names of species, just as I had taught his father, Rusty, each of our horses' names. Small groups of sanderlings ran on wet sand, chasing retreating waves, eating sand fleas. Godwits, willets, long-billed curlews, sandpipers, and semipalmated plovers strode ahead of us. Beyond the row of houses, where no people lived, pelicans sunned between feeding times, tucking their long gullets against their chests and facing out to sea.

Pelicans nest on two of the four northern channel islands in large, noisy rookeries. From the two or three eggs in each nest, stark-naked hatchlings appear—homely, awkward, and helpless. A medieval legend contends that the male pelican often kills its young but if the mother sits on her dead hatchlings long enough and regurgitates the blood of her dead, the young birds will come to life again. Maybe that's what Sam and I needed.

On days when we did walk, the birds flew up ahead of us. Once almost extinct from DDT, pelicans flew in a chopped-off V, clipping the crests of waves with their wing tips, flapping in a slow progression, each bird taking the communal rhythm from the bird ahead. When fishing, they can drop from a great height, reaching speeds of sixty miles per hour, their wings folded back into an arrow; and when they hit the water, their long bills act as "cutwaters" to reduce the blow. After catching a fish, they bring it to the surface and swallow it whole.

Sam's vitality returned. His brush with death had engendered my own leap in strength. If I was dragging around, how could I expect him to recover? On our morning outings he greeted other beach walkers, whether they wanted to be greeted or not. He'd grown up on a ranch, where every human he saw was a friend; how could he have understood the idea of a world so populous that there could be such a thing as a stranger?

One woman seemed delighted by his attentions. Smiling, she bent down to pet him: "I was just wondering what it would be like to see God when your dog ran up to me."

"He is a god, but spelled backwards," I replied.

chapter 18

At the ranch I'd kept my vet kit in a waterproof bag lined in green plastic. In it were bottles of penicillin, Combiotic, vitamins, handfuls of 16-gauge needles and various-sized syringes, a turkey baster to irrigate wounds, salves, creams, mineral oil, pinkeye medicine, Hoof Alive for quarter cracks on horses, and an aspirator to pull out mucus from the lungs of breech calves. The level of medicine we practiced was crude, but, then, our animals weren't sick very often and the survival rate was high, which I attributed to fresh straw and Mozart in the sun sheds rather than to my veterinary prowess. The animals were talked to, touched, loved. Now I wanted to better understand what makes people—or animals—live or die, the intermix of physiology and psychology, and to chart the dynamics of the patient-healer relationship. I wanted to look under the skin and see how the heart worked, how all the systems of the body—nervous and circulatory—achieved their miraculous harmony.

At 7:00 A.M. on a May morning I found myself in the Coronary Care Unit at Cottage Hospital—not as a patient this time, but an observer. I'd asked Blaine if I could go on rounds with him. I wore a white coat and carried a stethoscope. "Put on your roller skates," he said as I followed him from room to room, floor to floor. "Are you doing a rotation with Dr. Braniff?" one of the nurses asked. "No, I'm an imposter." "Oh, there's probably a lot of those around here," she said. Blaine scowled at her but he was laughing. We were "on the unit," in the room where I had been a patient.

When you are suddenly and acutely ill, you temporarily lose your self-image—no vision of the scenario in which you are involved. Everything is present tense: the gap is a place with no reference points—you float from one breath, one heartbeat to another. Looking at that patient hooked up to monitors, I couldn't believe it had once been me.

Blaine is an invasive cardiologist. He doesn't perform heart surgery, but he does put in pacemakers, defibrillators, and works in the heart catheter lab, where he performs angiograms and angioplasties.

"What my patients don't realize is that cardiology, as we know it, has only been in existence since the 1960s," he said as he pulled charts and went from room to room. We saw a woman with a pulmonary embolism, a man admitted with atrial fibrillation—a rapid heartbeat—another man with myocarditis—an inflammation of the heart sac—and a woman being prepped for an angiogram to determine if her chest pains were caused by coronary heart disease. With each patient he was gentle, gregarious, and full of humor—no deathbed solemnity for this man.

"It wasn't very long ago that we simply couldn't save the lives of people with serious heart problems. Now there are a variety of diagnostic tools and procedures that have demystified the workings of the heart. EKGs give us a record of the electrical activity of the heart—we can see when someone is having a heart attack; echocardiograms are sonograms that show us the heart muscle and valves in action—we can see how well or poorly it is working. When someone comes in with anginal chest pain, I can study their arteries by doing an angiogram—a simple procedure that allows us to find blockages if there are any."

Artificial valves, pacemakers, defibrillators, heart transplants, and bypass surgery came into existence in the early to late sixties, and angioplasties, during which obstructed arteries are opened up, thus avoiding open-heart surgery, only came into use around 1980. "We're still in the middle of a medical revolution and saving lives is routine around here," Blaine said.

But first things first. "First I have to discover what exactly is wrong with each patient. Medical students today don't spend enough time on simple diagnostic skills. They rely too heavily on technology. But when you have a whole bunch of symptoms and a complicated medical history, you have to listen and look and use your hands." In the nurses' lounge he poured two cups of coffee. "I can tell a lot by just looking at a patient. If someone's hair is coarse, it may indicate hypothyroidism; a diagonal crease in the earlobe means an increased risk of coronary artery disease; the angle of the fingernail and the skin around it may mean congenital heart or pulmonary disease. Palpating and taking pulses can tell a lot too."

The body is encoded. It is also an instrument inside of which the song of our lives is sung. As he hunched over an elderly patient and placed a stethoscope to the man's chest,

Blaine's eyes closed in deep concentration, as if listening to music. "Now it's your turn," he said and held the stethoscope so I could hear. What he was teaching me was the language of the heart, the notes of its percussion, and what each heart sound signified. The closings of the tricuspid and mitral valves make the normal heart sound: soft thuds. But the heart can gallop: there is the S4 gallop, made when the left ventricle is stiff and incoming blood hits the wall hard; and the "lub de dub" sound of the S3 gallop made when the heart's chambers have trouble ejecting blood.

"Can you hear it?" he asked, moving the stethoscope. But the sounds were faint. "The quality of each sound is important and so is the timing," he explained as we dove down a staircase to another floor, talking all the way. There are plucking clicks, high-pitched murmurs, or the soft blowing sound of a leaky valve. I heard the rubbing sound of an infected pericardial sac, and the harsh sound, "like water going through a kinked hose," when the aortic valve is too tight.

We visited a young woman who had had a heart attack while in Blaine's office and he had rushed her to the hospital where surgery was performed. She was mending well, but when we entered the room she started yelling at him, "I want to get out of here."

"You've had a heart attack, and if you weren't in here, you'd be dead. So take it easy for a few days, okay?" Blaine said.

"I'm going home now," she said. Blaine asked what was bothering her. "Nothing," she snapped.

In the hallway a lab technician delivered the woman's toxicology report: cocaine. "It brings them in here more often than

I'd like to think," Blaine said. "Cocaine can cause a sudden and severe vascular constriction until there's a complete blockage, and boom, that's it. Sudden death, or if you're lucky, you get to a hospital on time."

In the middle of rounds Blaine's beeper went off. "They love having me on a leash," he said, calling the heart catheter lab. His patient was being readied for her angioplasty. Earlier, an angiogram had been done to see inside the coronary arteries. A lesion was found. "Here are her pictures," he said, pulling a long wrinkled scroll from his coat pocket. The artery looked like link sausage where the obstruction pinched the vessel tight.

"This is what a heart attack-in-the-making looks like. It's her left anterior descending coronary artery. If there were a complete blockage, with no blood and oxygen getting through to the heart, the heart tissue would die in four hours. That's why it's so important to get to a hospital right away when you have chest pain. She still has a little blood getting through, and if we can get our balloon in there, she'll be home in two days. Otherwise, she'll have to go to surgery."

Blaine looked like Nehru in his starched cap, creased from front to back like a blade, and he wore a leaded vest belted over surgical blues. One of his partners joined him to assist. Even with a mask on, Blaine talked in a continual stream: consoling words to the patient, jokes with the nurses, explanations to me.

"The groin has been anesthetized and we insert an introducer sheath up the femoral artery, then a guiding catheter to the opening of the left anterior descending artery. It doesn't hurt." Blaine grappled with what looked like a mile of thin wire fed to him by a nurse. Up it went. A fluoroscopic camera was positioned above the patient's chest, and on the monitor I watched the progress.

"Now I'm going to try to get the wire through the lesion. See it? There it goes." I saw the narrowing inside the artery and the head of the wire poking through. "And over the wire, I introduce a tiny balloon. Now I'll try to get it right into the center of the obstruction . . . and by inflating it there, the plaque in the artery will be flattened against the side of the vessel, opening the artery again. There it is." To the nurses: "Okay, inflate."

A still picture was taken of the newly opened artery, which could be replayed on the monitor. Blaine asked to see it. "Looks like a good result." He leaned close to the woman's head. "It's all over. You're fine now."

Downstairs in the dictation room, Blaine recorded the findings from morning rounds as well as the results of the angioplasty. Dictations were rapid-fire, but during one of them, I heard him say: "Upon inspiration, no discomfort was experienced," the word *inspiration*, reminding me that even in this secular culture our language still lets us know the ways in which flesh and spirit are interfused.

An intake of breath is not just oxygen, a pulse is not just the rush of blood but also the taking in of divinity through an orifice, and as it moves through, it becomes a spark. To be inspired is to have accepted spirit in the lungs and heart, to watch it circulate through miles of blood vessels and capillaries whose tiny fenestrations allow oxygen, nutrients, and grace to leak into the tissues of muscle and consciousness, then be taken up again, reoxygenated, and returned.

"You've seen a lot of technology this morning but not the spiritual part of medicine. That's just as important," Blaine said. It was only nine-thirty and he still had to finish rounds before seeing his office patients. We shared a cup of coffee on the run as he described his next patient: "She has diabetes, heart problems, and she's in here with pneumonia. She's got a high fever and she's not responding to the antibiotics. We've tried everything, but her attitude isn't very good. She's had a hard life. I wish I could think of something to do for her . . . right now, her chances of surviving are about fifty-fifty and that's not good.

Blaine strode into the room calling her name: "Helen, it's me. . . . How are you feeling?" She rolled her head from side to side, an oxygen cannula in her nose. Blaine pushed his hands under her shoulders and lifted her heavy torso. "Helen, open your eyes, it's me, Blaine." Her eyes opened as he hugged her. "I'm awfully worried about you. You've got to help me out. You're not doing so well and we've got to get this temperature down together. Okay, Helen?"

She finally looked at him but didn't smile. "We've been through a hell of a lot together, and I know you're a fighter. . . ." Her head rolled to one side. A nurse came in and Blaine shot her a worried look. "Helen, keep trying and I'll be back this afternoon. If you want something that you're not getting just ask for it. You can have anything you want." He gave her a kiss on the cheek. "See you later."

In the hall he confessed he didn't think she would live. "We've tried every course of antibiotics there is and she keeps getting worse. There's no excuse to die of pneumonia at her age. I feel as if we're failing her somehow."

Halfway down the stairs, Blaine was beeped back up to the CCU. A patient of his, a man in his seventies, bright, articulate, but frightened, was on his way to radiology to have an arteriogram of his leg. The prognosis was bad: if his circulation didn't improve, his foot would be amputated.

The nurse lifted the blanket so Blaine could see: the foot was gray and cold, and on the verge of turning gangrenous. Blaine held the man's foot in his hand, then told me to touch it. "I don't want to have the damned foot cut off," the man said. When I looked up, Blaine was stroking the man's forehead. I had seen him touch the face of an older woman, but this was different—this was a rare glimpse of tenderness between two men.

"Don't worry, it might not be as bad as you think. . . . Just hang in there," Blaine said softly. He held the foot again, looking into his patient's eyes. "Come on, let's get some blood down there," he said to the man. "We've still got a day before you have to go to surgery. . . . Let's see what we can do." At that glimmer of hope, the man snapped out of his misery for a moment.

Blaine's beeper went off. "They're after me again," he told the man, then disappeared.

A half-hour passed. While in the ER, his beeper went off

again. It was one of the CCU nurses: "Come up here, Dr. Braniff. I think your patient's foot is getting warm." We ran upstairs. The blankets were laid back, exposing both of his feet. Blaine touched one, then the other, then I touched them. The gray foot was turning pink. "You've done it," Blaine said, then put his hand on the man's shoulder and shook it joyfully. "It's warm!"

"Bring him a big breakfast," he said to the nurses, then to the man: "I'm sending you home. You don't need to be here anymore."

We had coffee in the cafeteria. "What are you, some kind of witch doctor?" I asked, laughing.

"I didn't do anything. It was pure serendipity," Blaine said.

"You were tender with him."

"He was scared. The blood's not going to flow when you're feeling that way . . ."

"That's the point, isn't it? That's the chemistry of healing."

Upstairs we stopped at the nurse's station to look at Helen's chart. "Helen!" he yelled from the hall. "Why didn't you tell me it's your birthday today!" When he gave her a hug, she smiled. He called the nurses in. "I want to order her a special dinner—Greek food. Whatever she wants and a glass of wine—you want red or white? Make it two glasses. I'll come in and have one with you," he said. "And a cake. Don't count the candles—or do you want baklava? The sky's the limit." Helen laughed until tears came to her eyes.

▪ ▪ ▪

My thoughts about medicine were changing. How dynamic the human body is, the same dynamism as the ocean's, all the systems—circulatory, nervous, immune, endocrine—so vigorously interactive with the workings of the mind. From cell to psyche, there is a whole intelligence at work.

"Until recently," Blaine said. "No one quite understood the connection between the immune and nervous systems, but there are direct links." He explained that stress hormones from the adrenal gland can kill the brain cells that process short-term memory. "That's why I spend so much time on what we call attitude, which, when you look at it, is really a complex, mind-body phenomenon. I don't really believe in medical miracles. People should give themselves more credit for their healing abilities. A doctor participates in the process, that's all. One of the best things a doctor can do is encourage a tough, fighting spirit and a sense of humor. Those people almost always do better than the others."

In the morning we visited Helen. She was sitting up in bed with her hair combed, color in her face, drinking orange juice. Blaine looked at her chart: no temperature. "How are you feeling today?" he asked. "Fine," she said flatly, wondering what the fuss was about because she had no memory of how sick she had been. "You look wonderful," Blaine said, beaming.

It was a week of minor miracles, even though he didn't believe in them. His beeper went off. "Put on your roller skates," he said, flashing a smile. "This is what I love about cardiology," he said as he ran down three flights of stairs. "There's never a dull moment. I mean, my god, I could never have been a dermatologist."

In the emergency room he saw an older woman who had come in with a severe arrhythmia and tachycardia—a fast heartbeat. He listened to her heart with his eyes closed, then put his hand on her left shoulder. The expression on her face suddenly changed. "Look at the monitor," the nurse said. Blaine looked up. The woman's heart had converted to a normal rhythm. "There, feel better? You can go home now," he said.

In the hallway I asked if he converted heart rates often. After all, I'd seen him do a cardioversion with a defibrillator earlier in the week. "It only happens once in a while," he said. "It's unpredictable. Patients just need to be touched, that's all."

chapter 19

Does fire put out water? The smell of fire woke me and the eaves of my cottage framed maundering banners of smoke. I had heard how midwestern rivers can combust into rivers of flame, how gaseous slicks that look like floating galaxies can explode. Sometimes I imagined water was made out of flame, that rapids and waterfalls were incandescent coals, and that when fire trucks came their hoses spewed out sparks. But this was no house fire. A city—Los Angeles—was burning.

Balance that against the beauty of my long-legged friend Hillary swimming at dawn, her body curled inside a translucent wave, just her flippered feet waving, then her head bursting up in the midst of a school of dolphins and her loud yip of joy. For a moment I couldn't tell which fin was Hillary's, which a dolphin's.

By afternoon smoke hovered over water. Beneath, brown kelp was the rubble of burnt buildings. The surf was up, trying to vanquish what had already been destroyed. I sniffed the air and remembered the stench of my own body in the bathtub after being hit by lightning.

I couldn't sleep. It still startled me to be able to turn on a television and watch round-the-clock news after seventeen years of one or two fuzzy channels in Wyoming that went off the air at 10:00 P.M. On CNN I saw two handsome African-American men saving a Japanese man from being beaten by whites, black men beating up a caucasian truck driver, white cops bludgeoning a black man, and African-American rioters robbing Koreans. Every racial and cultural line was like a hot wire that had been crossed. One call from a friend in Los Angeles came through. After driving to his house across the city at midnight, hours after the mandatory curfew, he said: "No one stopped me, no one bothered me. It was like *On the Beach.* I think I was the only passenger car moving in the whole city."

In the morning, fog rolled in backwards over smoke, then folded into waves, closing them so they couldn't break, as if sealing in rage. A dead seal floated by with birds perched on its bloated side, picking at its flesh. Jim, who lived down the beach, ambled by with his usual cigarette and glass of lukewarm tea in his hand, because Dennis, his dog, was lonely for Sam. As they played, Jim scanned the smoky skies: "It's all those shelved screenplays of mine burning. It'll take years to get them all," he said sardonically. The rioting continued.

Blaine was on call that night and I decided to visit him. A long-time patient of his came in at 8:00 P.M. with a massive heart attack: a heart rate of 240 beats per minute and falling blood pressure. Because she had not called for help immediately when the pain began, she lost 50 percent of her heart muscle, since tissue death occurs when there is no blood profusion. An hour later, when I arrived at the hospital, she had already gone into cardiac arrest twenty times. "I've never had a patient arrest that many times in such a short period and survive," Blaine said. "And she's no spring chicken, either."

Small, light-haired, in her late seventies, Alice refused to give up, though very few people live if they've lost that much cardiac tissue. "Each time she arrested, she seemed to remain conscious: she couldn't speak but she looked me straight in the eye as I stood by the bed with the defibrillator paddles in my hands. I kept warning her, 'This is going to hurt a little,' and she'd give me a nod and a look that said, 'Do whatever it takes.' "

The first time, Blaine zapped her at the lowest wattage, which, he said, feels like someone punching you in the chest, but that didn't do the job, so he upped it to 150 watt-seconds. Still nothing; and then 200 watt-seconds, which feels like a horse kicking you in the chest, and finally her heart started. "In the twenty times I had to do this to her she kept smiling and never complained."

Blaine stayed with her all night taking wolf naps in the room next door. During that time she slid into somnolence because of the low oxygen content of her blood. "She responds better to you than to us," one of the nurses said. Though she appeared unresponsive, he talked to her. "You're doing better, Alice. . . . I'm right here," he said.

Blaine asked for a new blood panel: potassium, calcium, and magnesium levels were checked. When Alice became confused, Blaine suspected her sodium level was low, so they restricted her intake of fluids and she became coherent again. At midnight he called one of his partners and went over a checklist of things that could be done for her. No omission could be found. "She's only got about a thirty or forty percent chance of making it through the night," he said, "but she's got one hell of a spirit."

The beeper went off: a man had been admitted to the ER with rapid atrial fibrillation—an arrhythmia. As Blaine worked to convert his heart to a normal rhythm—first by touch, then with medication, the cardiac care unit called with the code "911," which means cardiac arrest. Blaine bounded up the stairs. It wasn't Alice but another man and Blaine proceeded with his twenty-first resuscitation of the evening.

He checked Alice and decided to put in a temporary pacemaker to override her quixotic heart. This procedure took half an hour, and afterward, Blaine lay down in the room next to Alice's, but no rest was forthcoming: there was another arrest in the ER. "Maybe dermatology isn't such a bad idea," he said, descending stairs to the ever-bright emergency room.

Alice slept quietly, but with breathing tubes and a Swan catheter in her neck, she hardly looked human. Blaine went home, slept an hour, showered, and dressed for a new day of rounds and office calls. He called me at seven: "Alice has come out of it," he said excitedly. When he'd stopped back at the hospital she was sitting up and smiled at him as he came in.

"Unbelievable," Blaine said later. "Is it really the same woman? She shouldn't even be alive. I thought I'd lose her last night. That's what I told her daughters. I worried so much. . . . God, I'm wrung out." When Blaine went in to see her she pulled him close and said, "I care so much for you." There were tears in his eyes. "I care for you too, Alice. You've made a remarkable, almost unbelievable turnaround." She was so well at the end of the week they sent her home.

Survival is as much a matter of grace as fight. The expression, "grace under pressure" implies the attainment of equanimity and equilibrium. The fundamental durability of the human body surprises us because the pain can be so intense—yet pain is often transient and hides the tremendous efforts the body is engaged in to heal itself.

We're almost always unaware of the interior adjustments being made: chemical changes, dead brain cells being carried away by scavenger cells, new dendrites forming with each new thought and memory, tissue healing, heating, and cooling, blood pressure and heart rate adjusting, kidneys and lungs interacting . . . It's not that this self-regulating system can't be overwhelmed and stop working, but so many variables make up the living and healing process, it is often impossible to know who will die and who won't, and why. Add to the biological mechanism the mix of doctor, nurse, and patient—another kind of chemistry—and the possibilities for disaster and miracles grow.

"A positive attitude is absolutely critical," Blaine said. "If I walk into the room of a heart attack patient and see fear in his or her eyes, I know my main job is to get rid of that fear. If there has been tissue damage, life or death can depend on the health of the peri-infarction zone—the area around the wall of the heart—which in Alice's case wasn't moving at all. Fear releases catecholamines, which increase heart and blood pressure rates, which in turn can kill off that fragile zone. Then there's nothing left."

Twenty-four hours later Alice was back with chest pain. They did what they could to make her comfortable, and almost immediately, she improved. Blaine went off call that afternoon. He badly needed rest. At fifty-six he's vigorous but not indefatigable. The last time he saw Alice she was sitting up in bed eating dinner. Late afternoon sun streamed in her window and her reddish gold hair shone. He told her she looked beautiful. That night she had another round of cardiac arrests. The doctor caring for her—not one of Blaine's group but a doctor who didn't know her—had no luck keeping her alive.

chapter 20

At the end of May, Kate, Clyde, Jim C., Hillary, Jim, and I had our final barbecue dinner. Steaks and snags on the coals, artichokes and avocado salad, fennel bread, and Australian cabernet, while the dogs stretched out in cool ice plant after hard play. In two days Kate would leave for Australia to see her mother, Clyde would go to San Francisco to stay with his son, Jim C. would begin rewrites on a screenplay, Hillary would begin her book of "sealosophy," her husband, Jim, would go out on his boat to dive for abalone and sea urchins, and I would finish packing up my house, the lease being up, and drive with Sam to Wyoming.

"It will never be the same down here," Kate said. By midsummer she and Clyde would move into their newly rebuilt house, and on my return in late August, I'd be moving to another part of town altogether, if and when I returned. The casual, dog-filled intimacy of that cove had been broken.

I counted down the days: four left. Each was more brilliant and balmy than the last. How could I leave? Sam hung his head over the seawall, scanning the beach for dog friends. Impatient with my packing, he finally took off, running the half a mile to Dennis's house, and together they ran up the beach to see me.

When the fog came it filled my veins like a narcotic. I dozed on the couch, surrounded by half-packed boxes. It had been almost a year since the accident, yet I didn't want to move; I didn't want to leave. I was still so tired. On June 13, the day before my departure, Blaine gave me a last blood pressure test. When I stood after lying flat, my pressure dropped twenty points, but my heart pumped valiantly—as it should—in order to deliver more blood to the brain. But at a price: I felt clammy and broke out in a cold sweat, and the old chest pain started up again. On the other hand, the fact that my heart was compensating for low blood pressure showed that I was getting well. In the morning I started on my journey home.

chapter 21

Mid-June. I was in new skin: no history, no past, no future, only the present becoming more of the present. . . . I drove past condor sanctuaries, up the eastern slope of the Sierra, past Manzanar, and the water-poor Owens Valley. Just beyond Bishop, it began to snow. Six hours into the trip, I was too exhausted to drive farther. I pulled off onto a wide sage flat, bundled up in a sleeping bag, and breathed in the piquant mix of sage and snow.

When I woke, lightning was breaking over my head, its branches tangled in clouds, and all across northern Nevada thunderstorms collapsed on top of me. I drove in a trance, each lightning stroke a sword driven through my back, pointing at some new spot on the map. Sometimes Sam and I huddled ridiculously in the middle of the seat by the side of the road. Lightning was hunting me down on this lonely Nevada highway. "What are we doing here?" I asked him nervously. During another storm, we drove recklessly on curving mountain roads with a Shostakovich symphony turned up full blast, wishing we were already dead and being ferried away to the underworld.

Halfway through the journey, I received news that one of my dearest friends had died in a car accident. Attending his funeral was my first act on returning to Wyoming. The ceremony was in the high school gym and five or six hundred people attended. Bruce had been a newspaperman all his life. He could build a story out of vapor; he also loved good champagne and well-bred horses.

The whole town was there in its ragtag splendor. A young woman sang his favorite song to the accompaniment of a tape played on a bright yellow ghetto blaster, and the bleachers in which we sat were alive with random snippets of gossip. I overheard someone say, upon seeing me walk in: "Hell, I thought she was dead too."

Everywhere I went people told me lightning stories: of seven cows lying dead in a circle around a lightning-struck tree; of five horses, all killed walking a fence line; of the pointed ends of rowels in spurs melting; of a horse with a white streak down his shoulder and across his belly where he was struck; of houses exploding; of ball lightning coming down the chimney and bouncing into a baby's crib; of sheepherders walking for days unable to remember who or where they were.

Afternoons, storms dominated. The sky was so dark we had to turn lights on. At Stan and Mary's ranch lightning struck the field where their son was haying; a park ranger was struck while opening a ranch gate, and in Big Timber, where I visited a friend who is an EMT, a storm broke out during the rodeo and for safekeeping she made me sit in the ambulance. Sometimes I got up in the middle of the night and watched lightning flash whole mountain ranges in and out of existence and tried to understand how such a vast and distant phenomenon could have entered me.

The ranch looked the same but was different. The pastures hadn't been cut for hay, no irrigating water was set in the meadows, beavers had built a domed house attached to the lake's tiny island I called Alcatraz, and in the process had cut down the fringe of willows where grackle nests once hung in protective safety over the water. It was wonderful to see the animals—cows and calves, horses, dogs. Rusty was deaf and and didn't recognize me. He walked sideways because his left hip had given out after seventeen years of working livestock, but his eyes were bright and peaceful. Sam and his sister, Yaki, and I walked up to the spot where we had all been hit, but there was nothing to see.

Chest pains and dizziness nagged at me: this was no longer my home and I knew I had to leave. I left Sam in the care of Jennifer, a young veterinary student who had worked for us during the summers, and, not daring to look back, drove away.

I felt like a river moving inside a river: I was moving but something else was rushing over top of me. There was too much to take in: the deep familiarity with a place where I had lived for so long and the detachment a year away brings. The rivers were layers of grief sliding, the love of open spaces being nudged under fallen logs, pressed flat against cutbanks and point bars. I felt as if I'd never left, and at the same time as if I could never come home.

I drove. Chest pains seemed linked to open spaces: the wider the land, the more intense the feeling of constriction. I felt vertigo, the lodgepole pines on the side of the road slanted in on me, and when lightning broke into them I could see they had already burned: this was Yellowstone Park. On mountain passes snow inched away from seams of grass like an eclipse pulling away from the sun. Because I had missed snow, I stood in it barefoot until my feet went numb. My face was a moon, its dark side turned to views I'd looked at for seventeen years, and still I went blind. I tried to move out of myself and go into other things but the taut wires of an aching heart pulled me back in again.

Near the Wind River mountains I stayed with friends who are in their seventies. Every evening after dinner we went for silent walks—no talking allowed—crossing the Green River, hiking up to a pond. In that silence the world began to take on life for me again. Afterward, I slept in a lone log cabin on a high, wide meadow, and at dawn the symphony of sandhill cranes set me upright in bed and I thought, "This is the first time I've been embraced since I was dead."

When sun burned the sound of birds away, I moved on. Storms came. Where lightning hit the ground I wondered if an X would appear, marking that place as an intersection where all lives crossed or were blown apart, or if it was a moving line running across a tilting, spinning world. Thunder worked like echolocation: it told me where I was, where to go next, and lightning was the lamp that showed the way.

I knew that the traveler must dissolve nostalgic threads of personal history and go ahead with no baggage, no determined route; that the so-called hero is one who has mastered her own dissolution; that she's not a conqueror but a surrenderer, she is "geography's ant." Was I a citizen of the underworld with a ferryman who had gone mad?

chapter 22

After Wyoming I continued north to Alaska. I had a job teaching on a schooner for a week, then planned to join three biologists and help them count seals. I was tired of people, conversation, noise. Even though I felt lousy, I was well enough to need a change.

The local plane stopped in at Wrangell, Ketchikan, and Petersburg. I read Derek Wolcott's *Omeros*, a St. Lucia version of Homer's *Odyssey*: "You ain't been nowhere," Seven Seas said, "you have seen nothing no matter how far you have travelled." As I traveled, the red ember in my back—my exit wound—started to burn.

"Mark you," the narrator continues, "he does not go, he sends his narrator; he plays tricks with time because there are two journeys in every odyssey, one on worried water, the other crouched and motionless, without noise."

I was the one on "worried water." For three weeks I lived on boats wearing yellow and black: foul-weather gear and tall rubber boots. The first boat was a halibut schooner built in 1926 in Tacoma, Washington, with a long deep keel that drew seven and half feet. Her planking was two and a half inches thick by twenty-six feet, clear fir with oak frames, and had been used as a tender on both coasts and for hauling hay in the San Juans.

The first day out, after engine repairs in Petersburg, a lone humpback whale slid between the stern of the boat and the *Umista*, a dinghy whose name means "the return of lost things." The whale's head was long and flat, her flippers a third of the length of her body, and as she moved by I waited for her singing to vibrate up through the old timbers of the boat.

Everywhere I looked, rivers poured out of island mountains and mist rode the backs of green and clear-cut forests down to long-legged straits. This was a place where bodies of water met bodies of thought, even though I tried not to think about meanings but face directly whatever it was that constituted my life.

I lived in rain and what southeastern Alaskans call "clear weather," which is drizzle. At least there was no lightning. The tops of hemlocks and Sitka spruces pigtailed into pointed green fortresses, as if tightened by green purse strings. Rocking in and out of squalls, we calculated the time and strength of tides, easing through narrow straits on slack water between tides, eyes glued to the loran and radar each time we lost sight of the mast in mist. I continued reading *Omeros:* ". . . the 'I' is the mast; a desk is a raft for one, foaming with paper, and dipping the beak of a pen in its foam." But I took no notes, nor did I make entries in my journal.

In the evenings the boat spun on its anchor and mist fell to its knees, raining directly into seawater. Trees grew on red buoys, bald eagles lifted out of dark trunks like white-steepled chapels, a raven ate a crab in the boat's crow's nest, and schools of herring, who sometimes migrate in rolled-up balls five or six inches thick, broad-jumped the incoming tide.

Hiking the steep trail of one island, we came on bear shit, purple from berries and still steaming and three tiny ponds at the top of the hill held blossoming water lilies, Alaska cotton, and bog orchids. On seeing these, Jonathan, the captain, said, "It's so beautiful it makes me feel faint," and we hoped he wouldn't see anything *that* beautiful while at the helm. On the way down the mountain, we ran into the mother bear and her two cubs, and squeezed by her apologetically as she sniffed our air nonchalantly, then thrust her paw into a thicket of bunchberry shrubs.

When we entered Nelson Bay, a seal circled us, but dove when we whistled hello. Two years earlier the schooner had dragged anchor during the night and gone aground. I slept alone on deck in rain, wedged between winches, wrapped in tarps—it was Hitchcockian—like sleeping in a torn curtain. During the night, Jonathan yelled in his sleep and I listened for metal dragging over rock, though I wasn't sure if it was something I would be able to hear. The anchor held. At dawn Jonathan, Kerry, and I winched in the chain, shortened the lines of the *Umista*, and got under way while the four other passengers slept. The *Crusader* pushed through clouds until there was a clearing: ahead, Baird's glacier was a stepladder of retreating ice that led to daylight.

From Sitka I flew to Juneau, where I met up with Brendan Kelly, a seal biologist whom I had visited in the Canadian High Arctic the year before, and two National Park Service biologists who were friends of his. We headed "up bay"—Glacier Bay—by skiff for one of the distant glaciers. Out of Bartlett's Cove, whales breached like blossoming flowers and porpoises guided us around a wide bend to the entrance of the bay. In five hours we moved from spruce forest to cottonwoods to ice. Halfway there, cottonwood seeds floated thickly on water, a hint of how the surface would look dotted with ice. Then the water went from blue to green—glacial flour. We were moving backwards in time, through what Brendan called "a textbook scene of succession," from climax forest to bare, uncolonized soil, and where glaciers had retreated, newly exposed ground "rebounded," rose up from the compression of heavy ice.

On the way we paddled kayaks. Rafts of ducks three thousand strong parted as we approached, the bones of my paddle catching a shoulder of sun. Held in both hands, the paddle was a sheaf of lightning, dipping, flashing, dipping. I was a poor woman's wounded Zeus—ignoring chest pain and dizziness—my only guide the pointed bow of the kayak.

As we approached, ducks broke into disparate nations, flapping through mist. Bears roamed the beaches of every island, rolling in tall grass, but still I knew I had "been nowhere" and "seen nothing." I was just gliding . . .

Back in the Park Service's fourteen-foot skiff, we continued up bay, then turned west into Johns Hopkins Inlet. As the frequency of icebergs increased, we were forced to slow to an idle, pushing ice floes aside—heavy as despair—with oars. The inlet, half a mile wide, felt narrower because of the scale of the mountains: ahead, Mount Fairweather loomed to 13,000 feet. On both sides shear cliffs topped by towering peaks shed their snow in avalanches, and charcoal waterfalls tumbled over black ice, scarring white rock. Then we came to where the universe was falling apart.

The glacier looked industrial, 10,000 years of dredged-up dirt swirled into fractured ice and recesses of blue. It calved almost continually in thundering amputations—thunder with no lightning for once—and the shock waves that radiated out from the fallen fragment displaced so much water, the whole inlet undulated.

We camped on granite ledges near the face of the glacier. In back of us were vertical cliffs leading to the avalanched slopes of Mount Fairweather. Below, thousands of seals hauled out on floating icebergs and dozed on their backs with flippers folded like hands, or lay on their sides with tail flippers curled up, sometimes five to a berg, though some pieces of ice were only large enough to accommodate one seal. Behind them the glacier crumbled, an intricate city whose façade of blue light kept collapsing in.

From my granite perch I looked down bay and saw how the charcoal waterfall had bored four holes through a wide ledge of black ice suspended between two promontories of rock—an elegant Italian fountain of black marble. Is there any beautiful thing we have made that has not appeared first in nature?

While Brendan and Beth started their seal count, I traveled the entire inlet with binoculars. After being on a boat for two weeks it felt as if the granite was lifting and dropping, the way lovemaking in a cabin on the edge of a Big Sur cliff had felt vertical—dropping down to crashing surf, then rising on careening layers of fog.

The bay was littered with ice, and reflections of snow on mountain peaks strobed across water. Sharp reports and sounds like gunfire emanated from shattering ice whose tiniest pieces looked like creatures from the Burgess shale: forms we've never seen before but from whose primeval shapes came the idea for fox, loon, swan, boat, plane, bear, dog, whale, and human. There were no eagles there, just ravens—the ones who brought light to the world, then stole it back.

For a moment I sensed death's presence though I didn't know why. Maybe it had to do with geography: here, at the head of an inlet of dark water was the beginning and end of the world, and I could go no farther, but only face the glacier's moving wall of black dirt and light.

Death is a dark thing but it is also an illumination. Light is the other side of the coin from death, but the same coin nevertheless. When we are close to death or come back from death, we see light and move toward it—whether it has to do with "seeing into" things or staying alive. Ritual death followed by resurrection stands for the death of ego. It is the hero's journey and the teacher's—like Jesus' and the Buddha's—as well as any shaman's or healer's. To become a shaman is to have experienced a strong calling, often marked by a bout with near-insanitv or severe illness first.

Some Eskimos say that compared to shamans, ordinary people are like houses with extinguished lamps: they are dark inside and do not attract the attention of the spirits. Their word *qa-*

maneq means both "lightning" and "illumination"—because in their culture physical and metaphysical phenomena are considered to be the same.

In the rain at Johns Hopkins Glacier we built a lean-to with oars and tarps and set up our humble kitchen as shock waves from newly calved icebergs broke at our feet. We collected drinking water from a trickling spring that descended through dwarf willows on the slope whose steep wall bound us to our ledge. It was good to be away from the mists and sorrows of the human world, with its big trees and heavy schooners and green nightmares. Here the raw face of the glacier dominated, its deeply crenellated top, black and rough like a city seen from above.

Our large sleeping tent, staked farther up the slope, was shaken by the glacier's detonations and echoing thunder all night. I wrote in my journal by flashlight: "I feel as if I were a fish feeding at the crumbling edge of the universe." I wanted to wear clothes made from that place—perhaps an auklet feather skirt—and sleep on the white fin of an orca.

One day Brendan and Beth scaled the cliff and disappeared. Just a hike, they said. Leaving me with an aching heart on my slip of granite and no way home. Perfect, I thought. The perfect place to die, not that anyone said I was going to. My calendar reminded me that it was August 6th, the first anniversary of my lightning strike, and also Blaine's birthday. When we discovered this coincidence, he said simply, "As the New Agers would say, we were destined to meet."

While waiting for Beth and Brendan to reappear, I read a

Hudson Bay legend, translated by my friend Howard Norman, about a blind boy and his aunt who was struck by lightning. "I've been lightning-struck and I've become an eater of rotten seal-flippers, fish eyes and bird throats. This is bad luck. Bad luck. Nephew, life has taken a turn for the worse. What's more I'm more thirsty than you can possible imagine." After she was struck the boy held a vigil beside her. Her hair had turned "stiff as feather quills," her skin deeply creased, she moaned in pain, and a fever moved about in her body, which only increased her thirst. "A lightning-struck person gets enormously thirsty. You have to keep a close watch because such a person will try to drink the entire sea," the boy's father warned him.

I lay a plastic bottle sideways in the spring and let it fill, then drank the water. I had been that thirsty after being struck, and months later often woke in the night, burning up with thirst. Perhaps this glacier would quench me. Another city of ice collapsed, sending a shock wave so big it almost wiped out our kitchen. I thought of another Eskimo shaman story: "When the bear of the glacier comes out he will devour you and make you a skeleton and you will die. But you will awaken and your clothes will come rushing to you."

I scanned the mountains for the dark shape of a grizzly. Too steep, I thought, and not enough to eat along the way to make it worthwhile. When the tide went out, all the seal-inhabited icebergs floated toward the bay, then back again with the swing of the tides. This was the inlet of devotion and transparency where illusion washed back and forth, a place that could teach me to see. Night came but it was not dark, only gray, and the face of the glacier turned bright, as if a huge slab of moon had been cut off and laid against the mountains.

My friends returned shortly before midnight, bruised and exhausted. They had been lost and had fallen. I was glad to see them but their self-inflicted drama didn't interest me and I went to bed. I heard groaning: very slowly a house-sized ice floe rose from under dark water. Its hulking bottom half was clogged by black dirt and rubble, but on top lay an elegant blue pyramid of ice, like something from the drawing board of I. M. Pei.

chapter 23

"I felt my left arm burning and falling. Then I saw it drop. I screamed, but no one could hear me. The current made me arch back like a fish in a frying pan. I was floating. The most terrible pain turned into something warm and peaceful. I thought I was talking to my father, then I heard my heart start beating. . . ."

This story was told at the Third Annual Lightning Strike and Electric Shock Conference in which I was a participant. Sixty-five of us, all survivors of damaging electric shock and direct lightning hits had gathered in Maggie Valley, North Carolina, to tell our stories. The year before, the conference had broken up for a day because the lightning storms were so violent. The doctors who had come as guest speakers found people under tables and beds, and had to dole out tranquilizers instead of giving their talks.

We convened in a closed-down dining room and bar, sipping Cokes we bought from a vending machine in the hall. A number of the participants were severely disabled—some endured missing limbs, or had the shakes, motor and speech problems. A woman who had been struck by lightning while hanging clothes out to dry had trouble walking, talking, even standing up. When I sat down, the others looked at my name tag and asked, "Lightning or electric shock?" "Lightning," I told them. "When?" they asked, and I said August '91, and they nodded knowingly.

The group was organized by Steve Marshburn, who had been hit at his drive-up bank-teller window by a lightning stroke from a storm seven miles away. "It came in via the speaker and traveled through my arm, hand, groin, leg, exiting from my right foot," he said. Steve was driven to the local doctor, who knew nothing about lightning injury, so he was sent on to a neurologist in Wilmington, Delaware, who did nothing but give him a prescription for the headaches that had begun since the strike. The pain all over his body increased, his eyes became sensitive to light, he suffered frequent urination, insomnia, and couldn't grip with his right hand. "I couldn't seem to find any doctor who understood. They thought I was making it up, or was crazy." Another doctor put him in traction for six weeks, after which he went back to work. But on his way home he had a car accident and subsequently needed back surgery. Meanwhile the headaches and body aches continued, and his throat, still partially paralyzed, caused him to choke easily on food. A few years later he had a radical prostatectomy after a malignant tumor was discovered, which he feels was caused by the strike.

After being put on 100 percent medical disability, Steve Marshburn and his wife, Joyce, formed Lightning Strike and Electric Shock International, in 1989. "I was permanently disabled," he said, "and during a long recuperation period I started writing to others who had been struck. Then we realized there were long-lasting after-effects and that people needed help. So little is known about electrocution. Now we can direct people to doctors who have knowledge and experience with electrical injury and give them a support group to relate to."

Harold Deal was struck by lightning while walking from his truck to the house. He was thrown over a fence and into his neighbor's yard, a distance of fifty feet. "It was as if I had stepped into a very soft white cotton ball. It was so bright I couldn't see anything at all." He said he couldn't move his body and his head felt like it had been pulled down behind his shoulder blades. In the next four days he lost thirty-eight pounds. He couldn't sleep or eat.

He's now known around his town of Lawson, Missouri, as "Weird Harold" because the lightning affected his thermostat—a result of the injury to his autonomic nervous system—and he can't feel cold; nor can he taste food or feel sensory pain, and any lacerations heal extremely quickly.

In 1989 Harold Deal and Steve Marshburn and his wife, Joyce, joined together and while Steve and Joyce took care of the administrative tasks, Harold became the guardian angel of lightning strike survivors. He corresponds with any victim of lightning or electric shock, often flying to their bedsides to reassure them that they aren't crazy, that the symptoms they have are shared by others, that proper medical treatment is obtainable and that they will heal.

Now an airline underwrites his trips. Harold flew to Texas to visit an eleven-year-old boy who had been hit by lightning and lay in a coma for four months. When the boy woke up, he couldn't remember anything of his life. He suffers total amnesia. "I had to kinda start all over," he told me at the conference. "And Harold helped a lot. I didn't know who I was, who my parents were, what I was like as a kid. It's been kind of weird."

I heard more stories. An African-American woman was introduced to me because we were both from California. She looked shaken. "I just flew in," she told me. "I hate flying. Since my accident I don't care to have *anything* to do with the sky." Then she leaned in close and whispered conspiratorily: "I used to be white before I was struck by lightning." Everyone laughed. She glanced around. "Look at us. . . . My, my—I don't know what I'm doing here. With all of us around, lightning's sure to be attracted to this motel."

A stocky young man from Maine asked if I suffered from depression. I said no and he said he did, that he was on all kinds of pills and even though he still had a job, some days he didn't know if he could go on. A big-boned blonde sat with us and agreed that nothing in her life was the same. "I used to drive in stock car races for police charities for fun on the weekends. Now I don't do that. My body doesn't belong to me anymore."

Rose was electrocuted when turning on a bank of fluorescent lights in a theater. The switch blew up in her hand. Once a professor, now she has difficulty remembering her own name.

Robert had no vital signs after he was struck while dismounting his motorcycle in a rainstorm. The driver of a passing Power and Light truck, who saw it happen, stopped and gave Robert CPR. When the medics arrived, CPR was continued but they could not get a blood pressure. Robert remembers hearing something—a woman singing. Then the rescuers saw her too:

she was wearing a black dress and held a Bible. She knelt down by Robert, touched one hand to his chest and the other to the earth. Just then, one of the medics yelled: "I have a blood pressure!" Robert survived. While telling me this story, he pulled out a watch. "My father gave me this," he said. "When I was hit the stem was welded to the case but it still works." Implying, apparently, that his time had not run out.

Dr. Englestatter, a neuropsychologist from Jacksonville, Florida, who had been called in on Steve Marshburn's case years before, began the morning session with talks about the psychological effects of lightning strike or electric shock. While discussing what he called "postelectrocution syndrome"—depression, anxiety, panic, memory deficits, hypervigilance, exaggerated startle response, profound fatigue, restlessness, insomnia, impotence, night terrors—a young woman in the back of the room broke out crying. I suddenly felt clammy, nauseated, light-headed. "I have to get out of here," I whispered to Rose. She calmly put her hand on mine and said, "Yes, but where are we?"

Dr. Hooshmand took the podium next. Dressed in white linen, and of East Indian descent, he was brazen, kind-hearted, outrageous. His speaking style was part Florida car salesman, part physician, and he paced as he spoke. "You all have life sentences and it's not easy, is it, especially when nobody understands. But I do. *None of this is your imagination!*" he yelled out and smiled. He went on to say that doctors don't know what to look for. If they believe you at all, they give tests that show nothing of the damage to the nervous system: CAT scans, EEGs and X-rays.

"They're looking for an elephant when what they should be looking for is tiny damaged nerves, and they don't know how." As a result, so many lightning victims are misunderstood. Often the cause of the most serious damage is undetectable.

"There is no such thing as 'psychosomatic.' It's *all* brain. Every doctor should have a picture of the temporal lobe hanging over his desk. And if he says, 'There's nothing wrong with you, it's all in your head,' he's absolutely right, he just doesn't know it. But you do. You're not crazy, baby!"

He described how and why electricity courses through our brains and why almost all of us have had neurological problems. "Electricity enters the body and always follows the path of least resistance, and that means arterial blood, which is very oxygenated. It follows the arteries to the heart and up through the brain. Your brain is seventy percent water. When electricity comes in, you are being used as a conduit, and it fries your brain no matter where on your body the lightning enters and exits—it always circulates through the brain. That's why some of you see light. It's nothing mystical: it's your temporal lobe lighting up. Nerves are like wet noodles; after electrocution they are more like cooked spaghetti."

Lightning can affect the medulla, cortex, frontal lobe, occipital lobe, and temporal lobe. It can affect blood pressure, heart, memory, motor control, sleep and dreaming, cholesterol and blood sugar, and the entire immune system, and it can cause epileptic seizures. "The trouble is," Dr. H. continued, "nervous function doesn't show up on CAT scans or X-rays or EEG's. Those tests they give you are *useless*! Electricity doesn't go to areas doctors are used to, so they just dismiss it."

He described the sensory damage that occurs in front of the heart and in the back, in the thoracic spinal region. "And it hurts, doesn't it?" he said. "And they don't know why." He described dizziness, buzzing in the ears, headaches, numbness in face and arms, vision and hearing problems, impotence, frequent urination, cataracts, seizures, cancer, and chronic pain. A participant stood and said, "My tongue feels like it's burning all the time." Dr. H. nodded. "There are eleven billion nerve cells and ten billion guardian cells and they've all been fried!"

Dr. Mary Cooper became interested in lightning victims during her work as an emergency room physician in a large Chicago hospital. When a patient came in who had been hit by lightning, she went to the literature to find out what treatment was recommended and found that no such literature existed. She ended up writing the book herself. It is still the only guide for the treatment of lightning injuries.

"I grew up in rural Indiana. My parents didn't go to college and I was made to feel that only a man could go to med school. I was majoring in biochemistry, and one day the dean suggested I apply to medical school. I guess that's all I needed—someone to tell me it was okay, and I've never regretted it."

Just back from a conference of physicists and meteorologists, Dr. Cooper gave statistics: "The most dangerous part of a storm is before it really starts, when the convection activity is strongest. That's when lightning is most frequent. How many of you were struck when there was blue sky over your head?" Most of us raised our hands. She continued. "Lightning in winter is very dangerous because of the increased moisture content in the atmosphere. There are about three hundred direct strikes on humans per year, with a thirty percent fatality rate, which is high as fatality rates go, but it means that more people are injured by lightning as are killed."

She showed a slide of a headline from a tabloid: LIGHTNING TURNS WOMAN INTO MAN. She grinned. "Now, I can see how that could happen the other way around . . ." Another slide: lightning hitting a graveyard in Toronto—the gravestones lit up, the cityscape behind. A participant stood: "I was hit by lightning as I was walking out of church one Sunday."

More statistics: lightning carries ten to thirty million volts and passes through you in one thousandth to one ten-thousandth of a second. It's not the voltage that matters as much as the length of time it spends in your body. Skin, if it's dry and clean, is a primary resistor to electricity, but if the skin is wet—from sweat or rain—the resistance goes to nothing. It becomes a conductor. Lightning causes spotty damage to the protective sheaths around nerves and sometimes the cell itself, and these heal imperfectly. So when an ordinary impulse comes traveling down a nerve, it hits these damaged areas and jumps track, misfires, or crossfires, causing pain.

She told stories: "Lightning entered the hole in the throat of a man who'd had a tracheotomy. A whole baseball team was knocked down by a lightning stroke; one of them went into cardiac arrest: he survived for a few days, then died. One kid who was hit had an orange ball of fire come out of his mouth, fly through the air and kill a cow, then bounce back and hit him in the chest, killing him too. Now we think that electricity can also go in through the nose and ears."

A list of lightning-prevention tactics followed: during a storm, don't use the phone; don't sit in the bathtub or wash dishes near an open window; don't stand by a tree, or continue playing golf, or sit in a metal canoe on the water, or use a fishing pole, or ride a horse. After, Dr. H. stood once again and gave a list of foods and drugs to be avoided if you have suffered an electrical injury; no alcohol, chocolate, hot dogs (nothing with nitrites), MSG, or sharp or strong cheese. No narcotics with benzodiazepine base, like Valium and Prozac.

After lunch there was a small auction. Rose leaned over to me: "Last year I won an extension cord for a door prize." We laughed until tears came. She looked fragile and tired and I said something about all of us needing some sleep. She told me that since her electrocution she has never slept through the night. There are always "night terrors" that wake her. "One night recently I woke up and didn't know what my name was. I looked over at my husband and I knew his name but not my own. I lay there for a long time but nothing came into my mind. Who was I? Finally I couldn't stand it anymore, so I got up, found my wallet, and looked at my driver's license and had to study my picture and my name. Once I knew who I was, I could sleep."

In my motel room facing the Smoky Mountains, I opened the windows wide. Across the road, mountains rose up. It had been raining. The moon drifted through rain clouds, and intermittent patches of dogwood lit up as if, here and there, the moon was touching down in the thick forest. I thought of those humans who had awakened after being hit and became shamans and healers, and wondered what this new life of mine would be, carved from a ruined body and a ruined marriage, and what special passageways I could hollow out as in a labyrinth of dead ends.

chapter 24

I dreamed that the shape of this book should be a convection cloud, a rising bubble swarming with up and down drafts of electricity, moisture, and air. Inside, the narrative would zigzag like lightning and the pages would be laid end to end to resemble a tree trunk, a channel down which fire suddenly flows. Once the book had been read, the top of the cloud would explode leaving the reader holding a burned shell.

chapter 25

After the lightning conference I left Sam at the ranch in Wyoming herding sheep because, I rationalized, one of us had to work for a living, and because herding dogs are easily bored. I began dreaming again. I slid from the top of a cresting swell down into a green trough . . . A tsunami crushed the roof of my beachhouse . . . I was on a ship filled with dogs stacked like crates of strawberries. . . .

My travels had taken me from Wyoming to Alaska to North Carolina to Wyoming. Now yellow leaves spotted Wyoming's cottonwood trees and snow began falling in the high country where herds of elk traveled secret paths through black timber, grazing at dusk and dawn in tiny "parks," open meadows, then hiding again. It was time to go back to California.

Halfway down a Wyoming highway I stopped, turned around, drove back north, stopped, and headed south again. The evening before, I had laid a map of the world on the floor of a friend's house and spun a beer bottle around on it. Where it stopped, I'd go: Africa, Asia, the Arctic, Idaho. But I was barely well enough to drive to the next state. "You better just come home," Blaine said when I called in to report on my health.

The highway meandered and I meandered off the highway, little junkets down forest service roads to obscure mountain slopes. It wasn't the same, traveling without Sam.

There was no right way to go, no airborne fragrance luring me. I wanted to commit my body to some gravityless place. But that's what I had been doing anyway; for a year I had been treading water in the gap—it was my swimming hole, my highway, my home away from home, and it looked like I would continue to do so. Turn back again or not, all routes led to limbo, were paved with the stuff of limbo.

I returned to a wet fall in drought-stricken Santa Barbara. "This is the longest El Niño we've had in fifty years," one meteorologist exclaimed. El Niños occur when winds along the equator fall off and warm air moves eastward from the western Pacific. Sea surface temperatures climb and moist air in the central Pacific disrupts the jet stream, and, therefore, the normal seasonal rainfall in Asia, the Americas, and Africa. With the appearance of El Niño, droughts in western North America, northern Mexico, and central South America ended, while they commenced in normally lush southern Africa, the Philippines, and northern Australia. Who knew how long it would stay that way?

In the foothills of California, streams and waterfalls appeared where none had existed before. Houses slid into a rising ocean, dams spilled, steelhead trout swam up what had been for fifteen years a dry river. Every few days the sun rose through broken venetian clouds, then those were swiped by rain. House-sized boulders slid, and flooded streets were axle-deep in rainwater. Plowing through one of those deep troughs late at night, a stroke of lightning came down close, spearing standing water—a gold thread I could climb, if only I had someplace to go.

Once inside the house, I stayed away from open windows and didn't answer the telephone. Everything was wet. Huddled in blankets in the middle of the floor, I read the words of an African Bushman: "I enter the earth. I go in at a place like a place where people drink water. I travel a long way, very far. When I emerge, I am already climbing. I'm climbing threads, the threads that lie over there in the south. I climb one and leave it, then I climb another one. Then I leave it and climb another. . . ." In my sleep I climbed crazy ladders and green, twisted lianas that swung me from one internal storm to the next, and I rolled on the coiled heads of kelp roots—sea brains—that bounced onto shore.

Water ran everywhere as if pumped, the venous return of streams pressing into the arterial sea. Rain moved mountains; softened earth rode stream water down past the hardwood of manzanita and black sage, past the edge of the continent into halibut waters, and I thought of a place in the mountains where water seeped, went inside that spring through the branching nerves of ocher earth until it reached an opening—dank, wet, closed—a cave where the ceiling rained down deer, and when I tried to kiss those rocks, every animal drawn there leapt out, free at last, and ran hard under cold starlight mixed with human rain.

■ ■ ■

Sam waited out El Niño lying on his back, paws in the air. "Why fight it?" he thought. Some afternoons we both lay that way, with our heads butted against the living room's long windows to feel the storm's percussive blasts.

Even when the rain stopped the land kept moving. Sandstone bluffs ran into the sea, streams carved deep creases into smooth treeless hills, and the stars slid in meteor showers straight across the sky. In other seasons it was not water that moved mountains but fire that peeled scrub oak and chaparral, and in the seasons between fire and flood, the earth's tectonic plates lifted and thrummed like the flukes of passing gray whales.

I wondered if growing up in such a place had fostered my nomadism: I designed furniture that pulled apart, folded, and broke down into neat stacks. Since arriving in California, I had moved four times and it looked as if I would move again. Was it the land running under my feet or my feet running over the land? Water—in the form of rain and waves—runneled up through me.

One morning I watched a coyote pick his way through a cattle pasture and slide down an eroded bluff to the beach. He chased a lizard, then a sand crab. Under a rock, he dug out a half a sandwich wedged there by a surfer and took it in his paws, pulled the bread from the plastic sack and trotted off, bits of avocado sliding from his jaws.

Humans are almost as adaptable as coyotes. With every move to a new house and every gain in health, I said to myself, "This isn't bad, I can live with this," though in a month or two I had already shed that skin. I began to understand the meaning of "life sentence": my old life had been erased in one-thousandth of a second and now I was trying to fly with clipped wings.

Another Christmas came and went and another New Year. More dark and violent rainstorms undulated across bare hills and fell into the sea. I longed for snow—still and bright—and flew to Wyoming. As soon as I arrived, there was a January thaw. The temperature rose from twenty below to forty above in one day. A dignified friend in his seventies was seen trudging through melting snowdrifts in camouflage coveralls, Sorrel packs on his feet, and carrying a bent umbrella. A front moved through and snow turned to rain. A rainbow rose out of the river that sliced the ranch in half. Heifer calves, kept over from the fall to be bred the following year, were found stuck in mud that had thawed, then refrozen.

The fur coats of the ranch horses were so thick I couldn't tell who was who, with long chin hair, thick ankles and tiny mustaches on their muzzles. "My little gangsters," I called them, and they descended on me as a group, searching the palms of my hands for grain.

On my birthday, friends cooked a glamorous meal with ingredients Federal Expressed in from a city. I could only think of the birthday parties I used to give my dogs, all of whom were born in August. They were attentive guests and the menu was simple: half a pound of hamburger per dog, grilled medium-rare over coals in a pit dug in the ground.

I stood outside my small cabin and looked northeast—in the direction of my old ranch. As the crow flies, it wasn't far, but there was a mountain range between us. How might I fly there? I wondered. To that sacred spot halfway up a towering mountain, beneath a waterfall and above a quarry of dinosaur bones? I stared a hole through that mountain—a hole in the wall, so to speak—and like eyes whose muscles have weakened and wander in the head, taking no direction from the conscious mind, I could not turn away.

Just when the thaw broke and snow began falling thickly, it was time to go back to California. In February I had to move again. The new house was twenty miles north of Santa Barbara where cattle grazed open slopes that bend down to the sea. A rocky beach came with it and a view of my favorite island, San Miguel. That week Sam was put on a plane bound for Los Angeles to spend the winter with me.

I arrived at the Continental Airlines freight dock at 10:00 P.M. just as his cage was being unloaded. It had been six months since I had seen him. He'd gained weight, muscled out, and had a full Wyoming winter coat that made him look less like a fox and more like a long-eared bear.

On the way up the coast, we ate fat hamburgers—his usual welcome-home treat—cooked to order in a Malibu bistro. As we drove up the highway I wondered if he remembered the smell of the sea and the sight of dolphins. At the new house he looked puzzled. "Yes, we've moved again," I told him, though this time I'd picked a place as much for him as for me. There was open country to walk in, a peace as sweeping as the vistas, and he could come and go at will.

A bellowing cow woke us. Sam shot off the bed, out the door, down the hill, into the ravine, and up the far ridge toward twenty pairs of black angus cows and their calves. Once there, he sat down and gazed back in the direction of the house—a long distance away—his look asking: Why aren't you on a horse, why aren't we moving these cattle somewhere? Which is what he knew was right, being a veteran of intensive—as opposed to continuous—grazing.

Despite his disappointment about the cattle, Sam liked the new house, and so did I. Behind us was a wide corridor of grazing preserve, oak savannah never to be developed, and behind that, steep mountains with a blue coat of oak brush, manzanita, ceanosis as thick as Sam's winter hair.

Between storms, heads rolled—the tangled roots of kelp, that is, which looked like heads. In fact, what I was seeing were not heads but feet, called haptera, of macrocystis, a seedless, sexless kelp that grows in ocean water two hundred feet deep—at a rate of two feet per day. The beach was a seaweed salad: bits of sea lettuce, rockweed, wing kelp, and sea palm were tossed together, wrapped with delicate strands of eelgrass.

Torn clouds—clouds left over from the weekend's rain—strode across the sky, trying to catch up with the storm on its way; incoming clouds were shaped like spare parts: elbows, foreheads, and insect legs, and in their fast-footed skudding, they showed how our fragmented passions can be pieced together as stories.

chapter 26

Blaine called to see if I wanted to go on rounds with him again. I thought for a moment. "I want to see a human heart. Actually see one beating."

"Okay." Then: "You don't faint at the sight of blood, do you?"

We met at seven in the morning at the hospital. He had cleared the way for me to watch open-heart surgery with Rick Westerman, a thoracic surgeon, not from a glassed-in observation room but standing at the surgeon's elbow. I changed into hospital pants and top. In the scrub room Blaine handed me booties and a cap, tied on my mask, then dropped his own on the floor. "That's why I never became a surgeon. I'd never get out of scrub! My patients would die waiting for me."

We entered the OR and I was introduced to the surgeon, Rick, the assisting surgeon, Charlie, Gary the anesthesiologist, the pump technician, the resident in surgery, and various nurses. Blaine dragged a stool out from under the table and placed it at the head of the operating table between Rick and Gary, then waved goodbye because he had patients to see.

Stepping up on the stool, I was a traveler, a Marco Polo who had arrived in a place so exotic, few had seen it before. I peered over the blue towel, clipped between two IV stands, that served as a curtain to protect the patient's head. The surgeons were watching. Would I faint, they wondered? But what I saw was so abstract, so colorful and jewel-like, I wanted only to see more.

All epidermal exteriors are nothing, mean nothing, their purpose only to hide the forbidden cities within. The patient's legs and torso were bound in transparent Mylar wrap—a surgical dressing coated with iodine—mummified and inert. Nothing drew my eyes to those exteriors, not even the curled penis, thighs, chest, or the articulation of toes or arch of feet, so unsexed was he. Legs splayed, belly protuberant—a convenient shelf for instruments tossed down—the body had no head, or appeared not to. Eyes, looks, conversation were only surface, the skin of personality. Whatever else came into my line of vision in that first instant was incidental. Nothing could train my eyes from the view.

I was a voyager. How did they get inside there? The room was cold. Steam rose from the opened cavity. I felt as if I had broken into a hidden cave and come upon rubies and sapphires. Looking past skin, red tissue, white bone, into a chest held open by a steel frame, I saw a beating heart.

The surgeon moved his gloved hand under the heart, lifting it carefully with a small wad of gauze. Only slightly bigger than my fist, it had a covering of yellow fat near the top, but below, it was red and gray with branching arteries. On either side, thick pink flaps veined with black, like Italian marble, swelled and flattened: those were the lungs.

Not a landscape but an organscape: so many moving parts and bright colors—blue, purple, yellow, red. They were marble-quarries, veined leaves, red pathways leading to dark recesses. Was this a clock whose works were made from precious stones?

Before the surgery Blaine and I had looked at the patient's angiogram. A tight blockage in his right coronary artery and two tangent lesions in the left coronary descending artery were causing the problem. Blockage on the left side is more serious because those arteries feed all the organs as well as the limbs. The lesions occurred at the top of the artery where it branches off from the aorta, which made his a dangerous and urgent case. "What we call a widowmaker," Blaine said. "These are the guys who get an obstruction and die before they make it to the hospital. But his ventricular function is great. See? It squeezes down hard, and his pumping action is good, so that cuts the risk factor of surgery to less than one percent, and that's as good as it gets."

The beating heart was both militant and gentle; whether rhythmical or arrhythmical, its movements were soft, even subtle, but its persistent redundancy sustained life.

Dr. Westerman separated the left mammary artery, which would be used as one of the bypass grafts, held up one end, cleaned if off, then laid it on a green towel that Charlie, the assistant surgeon, had prepared at the side of the patient's chest. "We're looking awfully Christmasy," Charlie said. Everyone laughed except for the younger surgeon, who had just completed his residency. He was all concentration as he took from the man's leg a vein that would be used to bypass one of the blockages. Held in the air, it looked like pasta—a long rubbery tube that was laid on the towel next to the mammary artery.

Then the arteries were suspended by blue guy wires above the beating heart. Or was I looking at the blue span of a bridge, or the skeleton of a modern structure built over a graceful ruin?

"We have to stop the heart now," Westerman told me. "Because it would be too hard to do all this sewing with it moving up and down." Such a procedure could only happen with the use of the heart-lung machine, also called a pump-oxygenator, which reroutes all but a small quantity of blood away from the heart while it is being worked on and delivers blood and oxygen to the rest of the body.

"Pump on?" Westerman asked.

"Pump on, sir," the technician retorted crisply.

A tube emerging from the patient's atrium pulsed with blood, bright and beautiful as neon.

"Watch this," Charlie announced, smiling. He poured a pitcher of ice water into the chest cavity, right over the heart. "We're cooling him down to a core temperature of about thirty so he won't need as much oxygen." Gary started an IV with ice water. A human's normal core temperature is thirty-seven degrees centigrade. On the monitor—which recorded heart rate, blood pressure, and body temperatures—I saw the man's T Core drop. At 30.6 he looked dead but his reflexes were working; as ice water poured in, goose bumps rose on his thighs.

As soon as Westerman was assured that the pump was working properly, the anesthesiologist stopped the ventilator, which did the breathing for the patient, and added a high dosage of potassium to the heart—what would be, without the aid of the heart-lung machine, a "lethal dose."

"You may want to see this," Westerman said coolly. I peered closer: the heart jiggled, bumped out of rhythm, then gently stopped. No violent cessation, only a small change. How simple death is.

"Is he really alive?" I asked. The surgeons smiled and nodded yes.

Westerman slid his hand under the stilled heart and lifted it. He cut a tiny hole in the right coronary artery, inserted a vein that had been removed from the leg and sewed it in place. Five minutes later he started on the left side, wedging the heart into position with another wad of gauze. For an organ that pumps 100,000 times a day, 700,000 times a week, it looked rather meek nestled in the surgeon's hand. As Westerman prepared to sew, Charlie poured more ice water over the heart, cooling it and cleaning it at the same time.

Suction tubes pulled the water back out of the cavity. Sewing was performed with a half-moon hook and a tiny filament of blue thread that glinted under the lights. Westerman looked up at me: "Now don't go telling people this is easy. . . . They'll start having their neighborhood seamstress doing it."

How odd that sewing is thought to be "woman's work" when surgeons, sailors, and cowboys sew too. Yet how many female thoracic surgeons are there? And if precision motor activities are thought to be performed better by women, why wouldn't they make better surgeons too?

At 9:25 A.M. the bypass grafts were attached. "Let's get him going," Charlie said. The line on the heart monitor was still flat. I peered over the curtain. What if the heart didn't start? How can they be so sure it will?

Ice water was stopped, the heat in the room was turned up, calcium was added to the IV to excite the cardiac muscle into contracting. At 9:25 the patient's T Core was 34.5; and nine minutes later it was 35.7, almost normal. The heart moved, but erratically, almost vibrating. "He's in ventricular fibrillation," Charlie said, which is what happens to a heart when struck by lightning.

The EKG showed an unmetered scrawl across the monitor. Ventricular fib, as it's called, is a heart in chaos, a heart whose twitchings are so uncoordinated, it cannot pump blood. Death follows. The patient had to die before he came back to life—like anyone on the hero's journey.

"It's nothing unusual," Gary assured me. "When the electrical activity of the heart is stopped and started again it gets confused. That's what's going on now." Westerman asked for the defibrillator paddles. These were much smaller than the ones used to resuscitate a human with skin intact. Touching the paddles lightly to the ventricle caused the heart to stop for a fraction of a second; it started again when the body's natural pacemaker, the SA node, initiated an electrical stimulus in a synchronized manner, so that a normal rhythm could take over again.

The flat line on the monitor reshaped itself, rising into steep peaks and ever-narrowing valleys as if a cartographer was at work behind the screen refiguring a piece of land never seen before, or a composer disciplining wild notes into a serial order. Tibetan medicine gives us a body that is wholly circulatory. There are four groups of veins and arteries: "*thog mar chags pa' rsta*" means "first appearing vein," as if it came into the body like an evening star. "*Sriid pa'i rata*" is "vein of the world," reminding us how intimately we are connected to the entire cosmos. "*Hbrel ba' i rsta*," "the vein of union," and the last, "*t'se gnas pa'y rsta*," or "life-sustaining vein," remind us that the body is a self-regulating universe. About these arterial pathways the Tibetans say: "They all meet in the heart and the mental activities and emotions go through the chamber of the heart which causes the heart to beat."

Near the end of the operation Blaine stepped up behind me. "The will to survive isn't purely psychological. You've just seen that. It's built into our cellular structure; it's intrinsic. The heart is completely motivated to maintain life. The muscle cells may survive four or five hours after a blockage, but brain cells die in four minutes. That's what I like about hearts. It's what saved your life. It wasn't me."

At 9:34 A.M. the slow, sure accordian of the breathing machine started moving up and down in a quiet rhythm, and blood was allowed to enter the heart again. I watched the pink marble of the lungs inflate-deflate-inflate and the heart move quietly. "Pump is off, sir," the technician said to Westerman. "Is the calcium in?" Westerman asked. "Yes sir, calcium in."

The rookie surgeon finished sewing up the leg where the vein had been extracted and Westerman cauterized the needle holes and oozing spots along the grafts. He looked up at me: "What makes this go so smoothly is the people I have around me. It's the whole team."

Charlie threaded heavy-gauge wire through the chest wall

on his side, then Westerman did the same on his. Facing each other over the man's chest they each took handfuls of wires and in one fluid movement, pulled up: as their hands moved together the chest closed, the jewelworks hidden from view, and this stranger's body was made whole again.

chapter 27

"Love is a wildness that has been falsely domesticated," my friend Pico Iyer said. I thought of the bypass patient's chest being closed, the message being: You can't see this wild place again, you can't witness this beauty. But the moon was hidden in there, and the sun, and neither of these would rise or set, and the birds that flew up out of it were planets and constellations because the chest was really an aviary, too full of fluttering, and when it was closed no avian life would be seen again.

The thoracic cavity must have been the place where human music began, the first rhythm was the beat of the heart, and after that initial thump, waltzes and nocturnes, preludes and tangos rang out, straight up through flesh and capillary, nerve ganglion and epidermal layer, resonating in sternum bone: it wasn't light that created the world but sound. And the sewing up of the man's chest was like the closing in of a house with roof and walls: Where would passion erupt? How could the spirit fly free?

■ ■ ■

It was spring: light, sound, and smell were making the world anew. Light brought bare hills into being, but orange blossom perfume and coats of pale grass clothed them. A mile below my house, highway traffic droned, the occasional downshifting truck, the hydraulic rupture of "jake-brakes," a bleating siren; perhaps those sounds were here before the world was made, and out of its steady hiss, birdsong and human talking came into being, and from the friction of song and words on air, more earth was created.

Was it the sonic boom of a missle being shot off at Vandenburg AFB that gathered waves in tight around El Capitan Point, or was it light trying to teach the old lessons about clarity, with its green transparencies serving up impermanence, then breaking like glass on shore?

Night came. A shadowy vapor disassembled the new world. Do we have to make it up fresh every morning? I wondered. All day a distant foghorn pushed against heat until it was night and the full moon in apogee—the closest it would be to earth for fifty years—could not be seen. Across an avocado orchard a single reading lamp was switched on—its globe stood for the moon—but fog completely obscured the house. Along the whole coast, only that one light shone.

Fog lapped new shorelines halfway up the mountains, then receded, exposing the human mess below—a car wreck, a bombing, a wedding, a birth, a heart attack. Fog moved out like a pale eyelid being pulled back, but was the patient dead? In the middle of the night the sky cleared: then I saw the moon, and it did seem close, all its mottled brightness puckered down to its south pole. Heavy-ended above me, my shoulders hunched. I thought if the moon fell toward earth, my head would rise up through it like a spike.

Spring. A gray swell became a green swell: voluptuous, vibrant hills appeared, the green slopes of a woman lying on her side, her hip sprouting with wild oats, her thigh smooth with redtop grass bent over by canyon winds and tufts of bear grass and beach rye sticking out from her head. "I cannot be weaned off the earth's long contours," Seamus Heaney wrote. Nor can I.

Sam and I walked a four-mile stretch from east to west between Dos Pueblos and El Capitan Beach. Dos Pueblos had once been the site of two large Chumash villages and El Capitan had been a small *ranchería* called Ahwawilashmu. Up the coast a few miles was the village of Qasil, now Refugio Beach; then came Point Conception, where dead souls began their journey to the afterworld.

The Chumash said: "All those who die follow the sun. Sun sees everything. . . . The sun rises in the east and goes to the west and all the spirits follow him. They leave their bodies. The sun reaches the door and enters and the souls enter too. When it is time for the sun to fulfill his duty he emerges and he lights abysses with his eye and all who are in the dusk are resurrected."

As fog moved to the mainland I heard a flock of birds fly over. They sounded like a dress rustling, a dress being unfastened and dropping to the floor. Fog came unpinned like hair. On the beach cliffs, great colonies of datura—jimson weed—with their white trumpet flowers, looked like brass bands. Their hallucinogenic liquid was one imbibed by the Chumash, burning away reason and false domestication the way sun burns fog, and once again I watched the world remake itself and squinted my eyes to read the text of the blaze.

Denuded of most sand by winter storms, beach rocks were encyclopedias of design: etched in white quartz I saw a spiral within a spiral and a circle within a circle that fell over the pointed end of gray stone, designs that looked like fast-spinning propellers, and Saturn's rings, tree branches swept by a river, and the branching forks of lightning, and pushed up against them like bumpers were piles of rubbery kelp that had rolled in at high tide.

Up the hill from the beach, green slopes turned to hair: wild mustard—whose seed was originally scattered by Spaniards as they rode north—rose up on stalks like stilts, growing two inches a day until the flowering tops reached far above my head, and Irish green gave way to yellow. Hundreds of acres of avocados blossomed and the oak trees in the hills behind my house tassled out. In the fall the acorn harvest would be good. On the ground was buckeye, poppy, owl clover, paintbrush, but after thirty inches of rain the grass came in so thick the wildflowers wouldn't find enough space to grow.

Along the road, yuccas sent up their long-stalked flowers—creamy white candles, and on my walks at night, if there was no moon they brightened my trail. But when the full moon came around again it shone a tapering path on seawater that led from El Capitan to Point Conception: a candle lit in the land of the dead.

chapter 28

The mustard held sway on the hills. At the beach, Sam lay in elliptical slices of shade cut out of dark air by returning swallows. In the curve of the coastline, a mother whale and her calf—California grays—fed and rested thirty yards from shore. After they left I walked up the hill through yellow alleys, under monumental bouquets that opened out into green parks, then tightened into black sage jungles with pom-poms of purple flowers skewered on long stalks, which, farther up, gave way to scattered oak trees.

Back on the beach, fourteen vultures were pecking holes in the side of a dead seal and they flew up as I approached, welding together in the sky like a single black cape. Was this the hood that would flop over my head and send me down underwater? A vulture's sense of smell, not sight, directs it to prey. I hoped I was not giving off a wrong signal.

It wasn't death that came to my door that afternoon but Sam in great distress. I had heard an odd noise in the house but ignored it. Then he appeared, jerking in convulsions. I flew to him as his eyes rolled back and he fell over on his side.

On ranches I'd helped resuscitate calves with mouth-to-mouth respiration; I clamped my hand across Sam's muzzle and leaned down to breathe into his mouth when his eyes opened and he gave me a puzzled look, that said: "What the hell are you doing now?"

Gathering him in my arms, I laid him on the seat of the pickup and careened down the winding road to the vet. By the time we arrived he was looking quite well. Subdued but bright-eyed, he must have wondered what the rush was all about. The vet checked him for poison, bowel obstruction, fever, and infection but he passed all those tests. I told her we'd both been struck by lightning and that many lightning survivors later suffered from epileptic seizures. She concurred. "That's how he's behaving. He'll be tired now but he won't remember the incident. He'll be fine."

That night he lay with his head on my lap during dinner at a friend's beach house. A cool breeze lapped the front porch as we drank wine and ate steak. Sam didn't want any steak, he only wanted to sleep. The sound of the ocean and the flat-handed leaves of a sycamore tree lulled him, would heal him. Since being struck by lightning he had become hypervigilant: even the sound of popcorn popping in the microwave sent him cowering to another room. I thought of the woman on the beach who thought he looked like a god. . . . He wasn't a god that night, only a mortal whose body had been ravished by Zeus. The seizure was eidetic—a physiological reenactment of electricity's chant echoing in cranial chambers. Later, a friend called and said, "I was just thumbing through a Japanese dictionary and saw that the radical for *god* is the same as the one for *lightning*."

■ ■ ■

In one last storm, the moon was overtaken by clouds, as if its filaments had been crushed, and lightning's stun gun brought the sea into view—momentary, spasmodic bursts of white. The wall that stood in front of a hotel was torn away and the unfinished living room of a house being built filled with breaking waves.

A friend who had gone surfing said, "It started raining and the sky and water turned white and I saw someone stand up on his board in the lightning and there was white light all around him. He looked like an X-ray riding the last wave." The Chumash, many of whom lived at the water's edge and paddled their thirty-foot canoes across the channel, said of lightning: "Beware, that is an element from the hand of a power that caused us to see the world."

chapter 29

A wave is a disturbance on the surface of a body of water, a kind of derangement. Waves are born when wind drags itself across calm water and the friction pinches it up into ripples and wavelets, which later become waves. Wind, the ever-present gardener, thins out the smaller, weaker ripples by pressing them into whitecaps and in a saga of bathrhythmic natural selection leaves the larger wavelets to grow. As these swells move out from the winds that raised them, they take on a voyager's shape: less steep and wind-resistant, their crests become rounded, and they travel in sets, or "trains," of similar period and height for hundreds, even thousands, of miles.

Most of us are so earthbound, so terracentric, we think of the continent as the centerpiece around whose edges oceans lap. But to a set of waves journeying across the Pacific, the sea is the central body into which the lithosphere rudely bumps. The life of a wave ends at the edge of the continent where water becomes shallow. As they approach the shore, their length suddenly decreases, and to compensate, the waves slow down and steepen. The shallow bottom refracts waves: they are bent, not by a twist of wind but by the shape of the ocean floor. For a moment the wave is a mirror image of underwater contours, then, as it moves into critically shallow water, its back is broken and the long-distance runner falls.

The gravitational influences of sun and moon drive the tides. A full moon pulls the waters into a bulge; and spring tides, when sun and moon are aligned with the earth, bring on the big waves; and big waves bring surfers. In spring the surf came up, and the report on the marine weather station brought surfers in droves. They migrated from Rincón to "Hammonds," to "Edwards," to El Cap, to Jalama, running down twisting paths to the beach with their boards.

If human beings are fire watchers they are also wave watchers: eight-to-ten-foot waves had been predicted. After the tide passed its negative low point, it came back in with a storm-driven fury. Migrating godwits and plovers waited out the bluster on beach rocks, but the surfers ran to meet the waves that had journeyed so far for their pleasure.

Slathering sand on their boards for better grip they leapt on and paddled into walls of water, grabbing their boards and rolling under the foam. Positioned far out, they waited, paddling, sometimes gaining, sometimes losing a swell that rolled under them, until finally they were propelled forward as if the whole ocean were just that one wave.

A wave is speed plus time equaling distance that has taken a particular shape. Surfers are destiny's warp, turning in the gyre, twisting time backward, paddling against current, wind and tide, then with it, turning back into the wave's curl. Surfers are acrobats of time.

Between sets of waves there is a gap, like the one between living and dying. The waves are time and surfers ride time's back. Then they glide over the gap, erasing limbo, but before reaching shore, they reverse their direction, as if turning back a clock, and meet waves head-on, where white water pushes down on their necks, releases them into a lull until the next set of waves.

To fall is to rise. . . . Random pressure fluctuations in the turbulent lower atmosphere create perturbations on the water and, in turn, these disturb air flow: all one current. Underwater we can't find where anything begins or ends, and up on top between sets of waves, surfers rise and fall, sliding around in the amniotic bardo of water that mirrors ground but is in no way solid.

A ringing phone woke me.

"Do you want to go to the islands?" Jim and Hillary asked.

"When?"

"In half an hour, and bring the lunch." Their boat is a twenty-five-foot Radon, designed and built in Santa Barbara for urchin and abalone divers. A small forward berth, hung with wet suits and flippers, opened onto a wheelhouse, with a broad deck behind where the air compressor is mounted on the stern and lengths of hose were coiled—air for the divers. Against the winch on top of the cabin, two surfboards were wedged, pointing toward the open water of the channel.

By eight-thirty we had passed the breakwater and glided in fog past bell buoys and channel markers. There was a northwesterly wind at twelve knots and a three-foot swell—choppy but not gut-wrenching. "See those little white clouds coming over the tops of the mountains?" Jim said, pointing over his shoulder. "We call those catpaws, or the fingers of death, because they mean twenty-knot winds in the outer waters."

We passed large groups of shearwaters resting mid-migration, and seals hooked their flippers to rafts of kelp and slept until we roared by. Halfway out we could see neither mainland nor islands. The sky was gray and the water was inkwell-green. We crossed a current line—one side was wrinkled and dark, the other, metallic and blue. Then the limestone cliffs of the western end of Santa Cruz loomed above us and the sky cleared.

What had been a three-foot northwesterly swell was now much larger, maybe eight or nine feet, and we slid down green slopes of unbroken waves into troughs that were all chop at the bottom because the current was going the other way. This was the infamous "Potato Patch" that had downed ships and boats much grander than ours. We were like a bar of soap bobbing, slamming down so hard we had to hang on to keep from hitting our heads. In the stern, Hillary's two dogs stood splay-footed, their bear-cub ears flattened, their black hair wet with spray.

Twenty-six miles out, between two islands, where the water was too deep to anchor, Jim tied the boat to a clump of seaweed—a kelp tie—the way we tied horses to sagebrush in country where there were no trees. Squadrons of pelicans and seagulls sat on each end of the beach and the backsides of breaking waves were aquamarine and turquoise.

We moved over a forest of kelp. "I want to dive here,"

Hillary said, squeezing into her pink and black wet suit. Tall, loud-voiced from deafness, she's a veteran scuba diver who has been in the waters off the Great Barrier Reef, all around Mexico, in the Caribbean, and in the southern and northern Pacific; and perched amid the diving and surfing gear at their house is her Steinway on which she's happy to serenade any visitor with Chopin nocturnes.

Looking down at the swirling canopy of kelp was to look at the top of a great forest. What lived beneath was much more interesting than the skin of the sea. A kelp forest houses and hosts hundreds of marine animals. Norris topsnails eat its fronds; crabs and lobsters use it as a ladder, eating invertebrates on the way; abalone feed on drift kelp—fronds that have been discarded—and bottom fish and schooling fish live in its shelter.

Hillary stepped into fins, mouthed the regulator, and tumbled backwards into the sea. As she swam away from the boat, Jim fed out air hose, then stopped. "That's enough. . . . She'll never come back if I keep giving it to her," he said gruffly, never taking his eyes off the water. "This is what my tender does all day, for hours and hours. He watches the hose, he watches for my bubbles. . . . You don't want a guy who's going to go to sleep on you when you're a hundred feet down with no air," Jim said.

The swell was sloppy. We rocked and rolled. Jim tugged on the air line, signaling Hillary to come back. Finally water bubbled at the stern and her head popped up. "God . . . I was down in a canyon and it was filled with bat rays . . . hundreds of them . . . five feet across, flapping all around me."

We motored to the lee side of the island. Jim eyed the surf as we went. "The waves out here are hollow tubes; they're scary and beautiful. We call the end section of these waves 'the toilet bowl' and try to make it through to the end without getting flushed," he said grinning.

The swell flattened and we dropped anchor near a beach. Jim pulled the surfboards off the cabin roof and threw them in the water. Hillary dove in, then Jim handed their two dogs, Skippy and Minke, to her and she placed them on one of the surfboards. Holding the board with one arm, she paddled alongside, and in the slack between sets of waves, the dogs rode to shore.

Jim was a pink arch arrowing into translucent water. Twenty feet down the bottom was visible—that's how clear it was. He emerged and saw me sitting on the rail in cutoffs and a T-shirt. "Well?" he asked.

"I left my string bikini in Wyoming," I said. I hadn't been in the ocean for thirty years but I wanted to go in; I wanted to blast the gray cocoon in which I had been suspended when my heart had stopped. "It won't hurt you," Jim said, laughing. I jumped. This wasn't a turbulent hell realm into which I was leaping, but the real sea with its china-blue elixir cushioning me.

Opening my eyes underwater, I was swimming in sparks. The shell of my body lifted off and was destroyed as cool water flowed in over new skin. Like a frog, I did the breast stroke, plowing despair aside. Pale blue poured in with bright light, as if, coincident with the theory of relativity, mass was exerting an influence on particles of light even though those particles were massless.

Earlier I had thought about a small Tantric scepter—a vajra—sent to me by a friend in Nepal. Small enough to fit in the palm of the hand, its five prongs stand for aggression, pride, passion, jealousy, and ignorance. Old friends. It's said that the sharp edges of the prongs are like razor blades: a reminder that we will cut ourselves if we live without awareness, precision, and basic sanity.

Vajra is associated with blue and with water, and people with vajra-like personalities are keen-minded and open to the multifacetedness of any experience—a diamond-like quality of mind that operates with unconditional clarity.

Bending my legs at the knee, I kicked out straight: the vajra spun, churning out diamonds until my head broke the surface of water. Up top, huge swells undulated, lifting and dropping, then slowed, steepened, and broke as waves. They frightened me. I bided my time for the lull, then, between sets, swam for shore.

On the deserted beach we cast off our clothes and rolled in hot sand because the breeze was cool. With the two surfboards we made shade for the black-haired dogs. Hillary took their collars off so they could be naked too. Soon we were hot and began to swim back to the boat for drinking water. My timing was off and the undertow too strong and a wave caught me. I looked around just as its crest dropped down on top of my head. As Jim would say, I was getting flushed.

Underwater I was lost again; it was like falling through white leaves. Earlier I had read a description of a neutron star's interior as being nuclear matter interleaved with sheets, strands, or droplets of quark matter, which is matter that has been compressed by extreme pressures. That's how my body felt; I feared I would never be able to breathe, and in my panic, gulped water.

My head broke through the surface, shooting up in foam. As children, Hillary and I had been in the same dance performance of "The Little Mermaid." Because she was tall and dark and I was small and blond, she was the witch and I was the mermaid. We did not remember each other when we re-met forty years later when I came across a photograph of the two of us—in costume on the lawn of my parents' house. Now, coughing up water and gasping for air, I saw Hillary facing me like a dancing partner, pulling my arm up as if to place it on her sun-darkened shoulder—the mermaid and the witch—and she was laughing hard.

At the end of the day we pulled anchor and headed home. Along the edge of the island, the pounding surf had carved out openings. In some of those caves, suns and planets and spider-handed humans had been painted by the people who had inhabited the islands for at least ten thousand years. It was too late in the day to go in; the surge, like a boulder rolling against the opening, would have trapped us. The dome of the cave looked like the bone-vault of a skull full of passing thoughts, the ocean filled and drained from it, the way blood fills and empties out of a ventricle. As we turned homeward, the Potato Patch jounced us hard. "Don't worry, there's only another hour or so of this," Jim said wryly. The swells were bigger than the boat. There was no view over them, and I wondered if there really was a shore.

chapter 30

June marked the end of spring on California's central coast and the beginning of five months of dormancy that often erupted in fire. Mustard's yellow robes had long since turned red, then brown. Fog and sun mixed to create haze. The land had rusted. The mountains, once blue-hued with young oaks and blooming ceanosis, were tan and gray. I walked across the fallen blossoms of five yucca plants: only the bare poles of their stems remained to mark where their lights had shone the way. I was still trying to get my bearings. My blood pressure had normalized but I groped along the path.

Walking had become an obsession. It was the way I moved in the world, achieved some rudimentary intimacy with a place. I had walked on almost every beach from Malibu to Big Sur, and though only in increments of four miles at a time, I did it often enough to parlay the distance into a thousand miles. My walks nearer to home were also vertical: up and down the mountain from beach to a ridge eyelashed with pines.

As I trudged, my feet planted themselves in vertigo. Where was I? Why was I there? No matter how far and often I walked, I was still living in exile from the ranching community that had been my home. When fog came, only the weather buoy off El Capitan, lost in grayness, offered a bearing—if only one of sound—its plaintive blast was the tip of my tongue touching I don't know what—the place where the bardo had been, which was now nothingness.

Early one morning I walked to the top of the road that led into the mountains. Far below, kelp beds and current lines carved the channel into a thousand watery islands. A hurricane off Baja sent warm air dripping—more like sweat than rain—and the storm surge lifted waves in fast-period jade panes. High on a slope of tall oat grass, a redtail hawk whistled impatiently as if trying to summon prey up into its talons, and a bobcat—one that had become quite friendly—played king of the road as I approached, then jumped sideways into chaparral. The Spaniards called the oak scrub that covers much of the southern and central coastal mountains chaparral, thus the "chaps" needed to protect the legs while riding these brushy hills.

Where marine air mixed with dry canyon winds, chaparral broke open into savannah and coastal live oak, *Quercus agrifolia*, thrived. Twin-trunked, stiff-armed, wizened, and venerable, the oaks were elephantine; their thick bark wrinkled at the crotch of each limb like skin. Mottled with pale gray lichens, wind-contorted, hung with green wisps of moss, they seemed to be the reservoirs of some ancient memory of this human-tormented part of the world, and from beneath, their canopy looked like a brain. These oak are judicial, their gray trunks leaning into a hill, balanced by a long arm that reached the other way, almost to the ground, then lifted up to suspend its green cloud of foliage.

Coastal live oaks have ancient origins. They have grown on the California coast for twenty million years, and their more ancient predecessors are very much like today's species. I picked up a leaf. It was tough and leathery, many-pointed and waxy—a botanic strategy to hold moisture during months without rain. Even the tiny stomata that cover the surface of the leaf like little mouths, letting carbon dioxide in and water vapor out, can shut down quickly in case of extreme heat or prolonged lack of rain.

Under those trees there was no wind. The green acorn pushed its pointed tip out of a stippled cup and grew hard. Each oak gathered stillness with its brawny arms and brightened the ground below: everywhere, oat grass ripened to the color of maize.

"Beware the ash, it courts the flash," one European legend warned, but oaks, on the other hand, called "thunder-trees," were known to protect against lightning. Said never to be struck by a thunderbolt—though in fact, they sometimes are—they provided a place of shelter during a storm. Even the branches or acorns lent safety to a house, and the wood from an oak that had been struck was hastily gathered as a charm against misfortune.

Oaks are sacred trees, trees of peace, marriage trees. They live for hundreds of years, and the felling of an oak is still deemed a sacrilege. In *The Natural History of Wiltshire*, a British naturalist wrote: "When an oake is felling, it gives a kind of shriek or groanes, that may be heard a mile off, as if it were the genius of the oake lamenting."

Marriages were performed under isolated oaks, and lovers who wanted to know if they would marry floated two acorns in a bowl of water. If the acorns drifted apart, one lover would be faithless or else the marriage would not occur, but if they moved close together, marriage was certain.

I'm not saying whether I floated any acorns, but the moon on water that night was a silver oak leaf folded on my tongue, and on it was written my fortune. Later, an eclipse—I watched it through open doors from my bed—covered those words, but I imagined myself as a tree pushing up through hard soil, or as a wanderer with a knapsack walking across the face of a blackened moon.

chapter 31

It was June again, almost two years since I was struck by lightning and I found myself packing for Wyoming—saddle, boots, slicker, hat, bedroll, and, of course, Sam. I had to smell sage and feel the crisp presence of autumn in summer winds. While I was packing the papers from my desk, a roadrunner appeared at the window of the living room. A dead lizard hung from his mouth. He moved from one window to the next, tapping, desperate to come inside. Earlier in the day I had heard his plaintive cries, a whining sound like that of a puppy tied up and abandoned. Through an open door he found his way into the room, and before I could stop him, he jumped onto my computer, lizard still dangling. I finally convinced him to leave. What did he want? I wondered. As I shooed him out through the French doors, he didn't fly—he ran, his long blue tail feathers a balance pole that swayed up and down, back and forth, both ends of the dead lizard bouncing.

I looked up "roadrunner" in *The Lives of Birds.* As part of a mating ritual, male roadrunners present food to a desired mate but won't relinquish the gift until after mating has taken place. I felt honored by the offer, but chose not to dine on lizard that night.

Friends met me at the beach, among them, Blaine. He talked about the heart's atrium—how, compared to other cardiac muscle, its muscle was "uncomplicated," by which he meant perfectly smooth, so that it could conduct electrical impulses more efficiently. That's how I wanted to be, a smooth conductor—not of electricity—I'd done that—but of whatever else blew in on the hurricane's breeze, so that, kneeling on the altar of sand and swimming in the ritual of tides, I could listen and watch and see and hear.

We sifted through rocks and shells, taking a sand dollar to commemorate the day. The sun had long since risen out of the sand dollar's hole and was getting ready to return. A night heron landed on a post, cocked one leg up until it disappeared into breast-feathers, then flew away. A month before, after a round of tests, Blaine had given me a clean bill of health, but some weeks later chest pains nagged me again. It seemed fitting enough. How could there be a certain end to anything?

Who could say of the Phoenix after it made its nest of sweet-smelling wood which the sun set on fire, and burned itself to ashes in those flames, that it then rose from them, it would not face more trials?

As we walked we came upon a shark's egg case, called a "mermaid's purse." The amber pouch looked like kelp, a flattened version of the gas-filled floats that raise the long stems of seaweed to sunlight. On each end, the tendrils that hook the egg case to a kelp bed for safety had lost their hold in the hurricane's strong surge and waved like tiny arms in the sea air. At the point where we were standing, a kingfisher flew in, perched on a rock, and peered down into incoming and outgoing water. Battalions of pelicans flew over in formation and a young seal, ready to molt, hauled out on the sand.

Blaine held the egg case against the sun. There's something lonely and appealing and unnerving about sharks. In the midst of gross biological and cultural mutations, nothing about the shark has changed in the last sixty million years: the corkscrew-shaped valve in their intestine, their rigid fins, the abrading denticulated structure of their skin. Because they lack an air bladder, which gives most fish their buoyancy, sharks have to swim about all the time, never sleeping, as if motion would untie the knot of evolutionary stillness.

Stillness and motion. How does the knot get loosened? In our tepid, human sea of constant change, the shark, in perpetual movement, represents immobility.

"Look," Blaine said, "I think there's a live shark in there." Then I saw it too: a miniature inside the rectangular pouch, bobbing in his teaspoon of fluid, so like the amniotic sea into which he would soon swim. "Let's give him a break," Blaine said, always fair-minded, and, walking out into a turquoise wave holding the tiny shark in his big hand, he let the mermaid's purse go.

Sam and I went to bed early so we could leave before dawn to avoid the desert heat. Between highway sounds I heard waves and thought how the curve of the coastline here had sheltered and nurtured live-born sharks, humans, and migrating whales. Here, at the edge of the continent, time and distance stopped; in the lull between sets of waves I could get a fresh start.

"Now you are sentenced to live," a neurologist at the lightning conference had told those of us who had been dead and revived. A sense of panic ensued, but panic is like fresh air. The world falls out from under us and we fly, we float, we skim mountains, and every draught we breathe is new. Exposed and raw, we are free to be lost, to ask questions. Otherwise we seize up and are paralyzed in self-righteousness, obsessed with our own perfection. If there is no death and regeneration, our virtues become empty shells. At best my virtues were small, but at least I could rely on panic. A carapace had been smashed by lightning and all the events that followed—divorce, loneliness, exile, and unmasking—had exposed new skin.

During the night I was awakened by the window, cranked open, rattling. Leaning over Sam's sleeping body, I looked out: an owl stared in at us. Owls have always been associated with death and night, with the "dead sun," the sun that has set and passes beneath us in darkness. That's why the appearance of an owl is thought to herald disaster, the other side of the coin from the phoenix who rises from the dead like daylight. But, in China, bronze vessels were made in the shape of owls and used as roof-finials, which were supposed to shelter the inhabitants from thunder and fire.

I peered up as the bird twisted his blunt head to look at me. He was, after all, just an owl who had found a convenient perch. I thought of the lighthouse at Point Conception and took the owl's presence for a middle-of-the-night, messenger-from-the-grave greeting. This was a dead man's wink, a lighthouse's watchful eye turning slowly, allowing me to see inside the ocean, the dark canyons where bat rays mated, the shaking seamounts where tsunamis are born, the perilous ledge where the continental shelf would someday break off, sending us who knows where.

In the morning a thick marine layer of fog that had smothered the coast for weeks broke open. When I put my spurs, snaffle bit, and saddle in the back of the pickup, Sam jumped in and would not leave again. Far up the coast, even in daylight, the Point Conception light still revolved its great head, and just before I turned inland, north toward towering ranges and oceanless basins of grass, I thought I saw that light wink as if to say, "Hell, yes, you're still alive."

To fall
is to return,
to fall is to rise.
To live is to have eyes in one's fingertips,
to touch the knot tied
by stillness and motion.
The art of love
—is it the art of dying?
To love
is to die and live again and die again;
it is liveliness.

—Octavio Paz